A Digital Twin Model for the Management of Indoor Environmental Quality in Developing Economies

This book presents Digital Twin (DT) in clear and accessible terms, showing how it can be used to improve indoor spaces such as homes, offices, and other occupied environments. It explains how DT can support healthier, more comfortable, and more energy-efficient indoor conditions through intelligent monitoring, analysis, and control. The book provides a practical, step-by-step guide to developing and using DT systems to manage key indoor environmental factors, including air quality, temperature, lighting, and related conditions that affect human well-being and building performance. It also addresses common challenges associated with adopting this technology and offers practical ways to overcome them. A unique strength of this book is that it conceptualises and develops a holistic framework for implementing DT for indoor environmental quality (IEQ) management, tailored specifically to the contextual realities of developing countries. By doing so, the book moves beyond general discussion and offers a relevant and actionable pathway for application. It equips academics, professionals, decision-makers, and interested readers with the knowledge needed to adopt and benefit from DT technology in IEQ management.

Olushola Akinshipe is a project manager and researcher in the built environment. He is currently a Research Fellow at the University of Johannesburg, Johannesburg, South Africa. He holds doctoral and master's degrees in Construction Management from the University of Johannesburg, and a bachelor's degree in Estate Management from the Federal University of Technology, Akure, Nigeria. He has published extensively, contributing to 40+ peer-reviewed journal articles and conference papers. His research interests include digital twin, digitalisation in the built environment, facility management, indoor environmental quality, sustainability, and construction project delivery.

Clinton Aigbavboa is a distinguished scholar with over 18 years of experience in the Built Environment and Construction sectors. He is a professor of Sustainable Construction Digitalisation and Construction Industry Research at the University of Johannesburg, Johannesburg, South Africa. He holds a

PhD in Engineering Management, a Master's in Construction Management, and certifications in Good Governance and Artificial Intelligence from the Massachusetts Institute of Technology. He has authored over 30 research books, co-authored six edited books, and published more than 1,000 accredited journal articles and conference papers. His research interests include Construction 4.0, smart cities, sustainable human development, and digitalisation in construction.

Routledge Research Collections for Construction in Developing Countries
Series Editors: Clinton Aigbavboa, Wellington Thwala, Chimay Anumba, David Edwards

A Building Information Modelling Maturity Model for Developing Countries
Samuel Adekunle, Clinton Aigbavboa, Obuks Ejohwomu, Didibhuku Wellington Thwala, Mahamadu Abdul-Majeed

Training-Within-Industry Job Programs for Improved Construction Safety
Lesiba Mollo, Fidelis Emuze and John Smallwood

Biomimicry and Sustainable Building Performance
A Nature-inspired Sustainability Guide for the Built Environment
Olusegan Oguntona and Clinton Aigbavboa

An Indoor Air Quality Management Framework for Municipal Buildings in Developing Economies
edited by Mpho Ndou, Clinton Ohis Aigbavboa, Wellington Didibhuku Thwala, and Abdul-Majeed Mahamadu

A Practical Research Process for Developing a Sustainable Built Environment in Emerging Economies
Andrew Ebekozien, Wellington Didibhuku Thwala, and Clinton Ohis Aigbavboa

An Adaptive Safety Incentives Framework for Construction Organisations in Developing Economies
Kunle Elizah Ogundipe, Babatunde Fatai Ogunbayo, and Clinton Aigbavboa

Regenerative Principles in Facilities Design and Urban Re-modelling in Developing Countries
Bukola Adejoke Adewale, Babatunde Fatai Ogunbayo, and Clinton Ohis Aigbavboa

A Digital Twin Model for the Management of Indoor Environmental Quality in Developing Economies
Olushola Akinshipe and Clinton Aigbavboa

A Digital Twin Model for the Management of Indoor Environmental Quality in Developing Economies

Olushola Akinshipe and Clinton Aigbavboa

Routledge
Taylor & Francis Group
LONDON AND NEW YORK

First published 2027
by Routledge
4 Park Square, Milton Park, Abingdon, Oxon OX14 4RN

and by Routledge
605 Third Avenue, New York, NY 10158

Routledge is an imprint of the Taylor & Francis Group, an informa business

British Library Cataloguing-in-Publication Data
A catalogue record for this book is available from the British Library

ISBN: 978-1-041-23389-3 (hbk)
ISBN: 978-1-041-24735-7 (pbk)
ISBN: 978-1-003-74333-0 (ebk)

DOI: 10.1201/9781003743330

Typeset in Times New Roman
by SPi Technologies India Pvt Ltd (Straive)

Contents

Illustrations

Figures

Table

Preface

The built environment is undergoing significant transformation worldwide through the integration of emerging digital technologies. Among these, Digital Twin (DT) technology has emerged as one of the most promising innovations, with the capacity to reshape how buildings and facilities are monitored, managed, and optimised. This potential is especially important for Indoor Environmental Quality (IEQ), which describes the condition of indoor spaces in terms of air quality, lighting, thermal comfort, and acoustics. IEQ plays a vital role in human health, well-being, productivity, and comfort, particularly because people spend more than 90% of their time within enclosed environments such as homes, offices, schools, and other buildings.

Despite the growing global interest in DT, its application to IEQ management in developing countries remains limited. There is still a shortage of context-sensitive frameworks and practical models that can guide implementation within the environmental, economic, and regulatory realities of these regions. This book addresses that gap by presenting a comprehensive, structured approach to using DT for IEQ management in developing economies. It provides an in-depth discussion of DT, its enabling technologies, cross-industry applications, and relevance to the management of indoor environments. More importantly, it introduces a practical framework and roadmap for integrating DT into IEQ management, thereby offering an actionable pathway for addressing implementation challenges while maximising the benefits of the technology.

This book makes an important contribution to both the theoretical and practical dimensions of IEQ management. From a theoretical perspective, it establishes a foundation for understanding how DT can be applied to monitoring and managing indoor environmental conditions, drawing on insights from multiple industries and existing DT applications. From a practical perspective, it presents a Management Model for indoor environments to support a dynamic, automated, and intelligent system for maintaining healthier, more comfortable, and more energy-efficient indoor spaces. In doing so, the book advances knowledge on the relationship between DT and IEQ management while also providing a platform for future innovation in the built environment. Its novelty lies in its holistic and context-specific approach, which enhances its relevance to academics, built-environment professionals, industry practitioners, and policymakers seeking to apply DT technology to support sustainable development and improve indoor environmental management especially in developing economies.

1 Introducing New-Age Technologies for the Indoor Environment

1.1 Introduction

The indoor environment has become an increasingly important concern in the built environment because modern life is predominantly experienced indoors. Studies consistently highlight that people spend more than 90% of their time within enclosed spaces such as homes, offices, schools, hospitals, and other facilities (Popkin, 1999; Klepeis et al., 2001; Cai et al., 2023). Consequently, the conditions within these spaces directly impact human health, comfort, well-being, and productivity. The indoor environment can therefore no longer be viewed as a passive backdrop for daily activities, but instead as a dynamic and influential component of human life. Factors such as indoor air quality, thermal comfort, lighting, acoustics, and ergonomics collectively shape occupants' experiences and determine the overall quality of indoor spaces (Dykes and Baird, 2013; Frontczak and Wargocki, 2011).

Indoor environmental quality (IEQ) has consequently become a vital issue in the design, operation, and management of buildings. A healthy and comfortable indoor environment is crucial for enhancing satisfaction, supporting performance, and safeguarding occupant's health (Altomonte et al., 2019; Andargie et al., 2019). However, IEQ is not determined by a single variable. It is the outcome of the interaction between building design, construction materials, ventilation systems, maintenance practices, climatic conditions, and occupant behaviour (Devitofrancesco et al., 2019; Altomonte et al., 2019). This makes IEQ management a complex task that requires continuous attention, monitoring, and informed decision-making throughout the building lifecycle.

The challenge of maintaining IEQ is even greater in developing economies, where rapid urbanisation, infrastructure gaps, and uneven levels of technological progress often affect the quality and management of buildings. In many of these contexts, issues relating to air pollution, environmental stress, and limited building performance management systems make the need for improved indoor environment management more urgent (Agbo et al., 2021; Garland et al., 2021; Matandirotya et al., 2022). At the same time, construction and facility management sectors in developing economies are often characterised by fragmented information systems and low levels of digital adoption (Aghimien et al., 2018;

DOI: 10.1201/9781003743330-1

Adeniran and Shakantu, 2023). These limitations highlight the importance of more intelligent and integrated approaches to building management.

Recent advances in digitalisation have opened up opportunities to improve how buildings are managed and how indoor conditions are monitored. Information and communication technologies, data-driven tools, and digitally enabled workflows now facilitate more efficient information processing, communication, and decision-making in building and facility management (Aziz, Nawawi and Ariff, 2016b; Araszkiewicz, 2017; Støre-Valen, 2019). Emerging technologies such as artificial intelligence, the Internet of Things (IoT), simulation, cloud computing, and cyber-physical systems (CPS) have laid the foundation for more responsive and intelligent built environments (Lasi et al., 2014; Botkina et al., 2018; Alaloul et al., 2020).

Among these technologies, Digital Twin (DT) has gained growing attention for its ability to create a virtual representation of a physical asset using real-time data. In the built environment, this allows for monitoring, analysis, and optimisation of indoor conditions in more proactive, data-driven ways (Rosen et al., 2015; Tao and Qi, 2019; Cai et al., 2023). For IEQ management, DT provides a means of connecting physical indoor conditions to virtual analysis, thereby enabling more timely interventions and improved operational efficiency. This chapter, therefore, introduces new-age technologies for the indoor environment and establishes the foundation for understanding the role of DT in managing IEQ in developing economies.

1.2 The Indoor Environment

The indoor environment refers to the conditions and characteristics present within enclosed spaces where people live, work, learn, heal, and carry out everyday activities. These spaces include residential buildings, offices, schools, hospitals, and other occupied facilities. In modern society, the indoor environment has become increasingly important because people spend most of their time indoors. Studies have consistently shown that humans spend more than 90% of their time in enclosed environments, making the quality of indoor spaces a major determinant of health, comfort, and overall well-being (Popkin, 1999; Klepeis et al., 2001; Cai et al., 2023). The indoor environment is therefore not merely the physical enclosure of a building, but the actual environmental conditions experienced by occupants on a daily basis.

The quality of the indoor environment is generally understood through the concept of IEQ, which reflects the overall condition of indoor spaces from the perspective of human comfort, satisfaction, health, and performance. IEQ is affected by a combination of factors, especially indoor air quality, thermal comfort, lighting quality, and acoustic quality (Dykes and Baird, 2013; Sakhare and Ralegaonkar, 2014; Wei et al., 2020). These dimensions do not operate in isolation; rather, they interact continuously to shape how occupants perceive and respond to indoor conditions. Frontczak and Wargocki (2011) noted that human comfort in indoor environments is affected by numerous interrelated environmental variables, while Heinzerling et al. (2013) emphasised that IEQ assessment must consider the combined effects of these variables to fully understand the true condition of the indoor environment.

Indoor air quality remains one of the most critical aspects of the indoor environment because of its direct impact on respiratory health, comfort, and exposure to pollutants. Pollutants such as particulate matter, volatile organic compounds, carbon dioxide, biological contaminants, and chemical emissions from building materials can adversely affect occupants when control measures are inadequate (World Health Organization, 2010; Al-Hemoud et al., 2018; Piasecki, Kostyrko and Goljan, 2021). Research has shown that poor indoor air quality can contribute to discomfort, diminished cognitive performance, respiratory symptoms, and sick building syndrome (Redlich, Sparer and Cullen, 1997; Burge, 2004; Ismail et al., 2022). Thermal comfort is equally essential, as temperature, humidity, and air movement significantly influence how indoor spaces are experienced (Fanger, 1970; ISO 7730, 2005; Khovalyg et al., 2020). Lighting and acoustics also play important roles, affecting visual comfort, concentration, mood, psychological responses, and overall physical and mental well-being (Galasiu and Veitch, 2006; Küller et al., 2007; Stansfeld and Matheson, 2003).

A healthy and comfortable indoor environment is essential because it influences not only the physical health of occupants but also their satisfaction, productivity, and performance. In office environments, better indoor conditions have been linked to improved work performance and cognitive outcomes (Allen et al., 2016; Seppänen and Fisk, 2006). In educational buildings, the quality of ventilation, temperature, and overall indoor conditions has been linked to students' health and academic success (Bakó-Biró et al., 2012; Haverinen-Shaughnessy and Shaughnessy, 2015; Turunen et al., 2014). Similarly, healthcare and residential buildings require careful attention to indoor conditions because vulnerable users may be especially affected by poor environmental quality (Nimlyat, 2018; Andargie, Touchie and O'Brien, 2019). This shows that the indoor environment is not a secondary concern in building performance but a central indicator that reflects how effectively a building supports its occupants.

1.3 Building Management: Information-Based Decision-Making

Effective building management is essential for ensuring that buildings continue to perform in ways that support their intended functions, user needs, and long-term value. Buildings are not static products but dynamic assets that require continuous oversight, adjustments, and interventions throughout their lifecycle. From operations and maintenance to refurbishment and adaptation, a building's performance largely depends on the quality of decisions made by those responsible for its management. These decisions influence the physical condition, functionality, environmental performance, and overall user experience of indoor spaces. Building management should therefore be seen not merely as an administrative task but as a strategic and operational process through which building performance is sustained and enhanced over time (Noor and Pitt, 2009; Araszkiewicz, 2017; Fennimore, 2014).

At the core of effective building management lies decision-making. Every effort to preserve, upgrade, operate, or improve a building requires appropriate responses to changing conditions and emerging needs. Such decisions may pertain to maintenance planning, environmental control, space utilisation, system adjustments, and performance enhancement. Araszkiewicz (2017) noted that building and facility management requires informed choices to preserve and maintain the functionality and value of built assets throughout their operational lifespan. This is crucial because poor decisions often build up over time, leading to inefficiencies in building performance, service disruptions, higher operational costs, reduced occupant satisfaction, and declining IEQ. For this reason, building management depends not only on technical competence but also on access to reliable information that can guide timely and suitable interventions.

Information is therefore essential for sound decision-making in building management. Buildings and facilities generate large amounts of information throughout their lifecycle, including design records, maintenance histories, inspection reports, operational data, user feedback, and environmental measurements. This information is generated by various built-asset stakeholders, such as designers, contractors, owners, facility managers, maintenance teams, and occupants. When it is properly captured, organised, and communicated, it offers a clearer understanding of building conditions and enables more effective decisions. Noor and Pitt (2009) argued that efficient facility management relies on integrating data and information from multiple sources, enabling managers to respond to operational demands in a coordinated and informed manner.

The flow of information among stakeholders is equally important. Buildings are managed by a network of individuals and groups with different roles and responsibilities. When information does not flow effectively among all stakeholders, decision-making becomes fragmented and usually ends up being reactive. Araszkiewicz (2017) observed that facility management objectives can be achieved when reliable information is effectively communicated among building stakeholders and personnel. Effective communication ensures that issues are correctly identified, priorities are understood, and responses are coordinated. This may occur through meetings, maintenance reports, written records, digital platforms, asset databases, and sensor-based systems (Ko, Pan and Chiou, 2013; Pishdad-Bozorgi et al., 2018).

The growing complexity of buildings has further reinforced the need for better information management. Modern buildings often feature interconnected mechanical, electrical, digital, and environmental systems that require coordinated oversight. In such environments, manual and fragmented approaches are no longer sufficient. Advances in ICT have created new opportunities for improving facility management by enhancing the collection, processing, and utilisation of building information (Aziz, Nawawi and Ariff, 2016a;

Aziz, Nawawi and Ariff, 2016b). Building management is increasingly defined by the ability to gather, interpret, share, and act on relevant information. Consequently, information-based decision-making underpins more proactive, coordinated, and intelligent management practices within the built environment.

1.4 Emerging Technologies in the Built Environment

The built environment has, over the years, undergone significant transformation driven by advances in digitalisation and information-based technologies. Traditionally, building design, construction, and facility management were characterised by fragmented processes, delayed information exchange, and reactive operational practices. However, recent developments in ICT have significantly changed how buildings are conceived, delivered, operated, and maintained. These innovations have introduced new-age tools and systems that improve information processing, coordination, simulation, forecasting, and performance management across the building lifecycle (Aziz, Nawawi and Ariff, 2016b; Støre-Valen, 2019). As a result, emerging technologies are becoming increasingly integral to how built-environment assets are managed, particularly in the pursuit of more efficient, intelligent, and responsive buildings.

In today's world, buildings are continuously evolving and becoming increasingly complex. And this complexity has made emerging technologies increasingly relevant in the built environment. Modern buildings are expected to provide shelter for occupants; in addition, they must also support and enhance occupants' health, comfort, sustainability, operational efficiency, and user satisfaction. Meeting these expectations requires improved methods of collecting, interpreting, and applying information in decision-making. Woodhead, Stephenson, and Morrey (2018) noted that digital construction is gradually moving beyond isolated technological applications towards more interconnected ecosystems of innovative technologies and intelligent systems. Similarly, Hossain and Nadeem (2019) observed that the digitisation of the construction industry reflects a broader shift towards Construction 4.0, in which data, automation, connectivity, and intelligent technologies are integrated into the built environment. This transformation has laid the foundation for several emerging technologies, including artificial intelligence, three-dimensional systems, cloud computing, simulation, IoT, CPS, and DT technology.

Among these technologies, Building Information Modelling (BIM) has played a vital role in laying the groundwork for digital transformation in the built environment. BIM allows for the digital representation of building elements and project data, thereby enhancing coordination, visualisation, and lifecycle data management (Garber, 2014; Kunz and Fischer, 2020). Although BIM alone does not create a fully intelligent operational system, it has significantly contributed to the shift towards digital thinking in building delivery and management. Other technologies, such as IoT and sensor-based systems, have further advanced this transformation by enabling real-time data collection from physical building components and operational environments.

Through interconnected devices and communication networks, it is now possible to capture and gather information on environmental conditions, system performance, energy consumption, occupancy, and equipment status in real time (Botkina et al., 2018; Alaloul et al., 2020). Cloud computing and simulation technologies have further expanded these possibilities by enabling large datasets to be stored, processed, analysed, and modelled to support more informed and predictive decision-making.

These global technological advances have greatly contributed to the development of CPS, which marks one of the most notable conceptual shifts in modern building and infrastructure management. CPS involves integrating physical entities with computational systems, communication networks, sensors, and actuators, enabling continuous interaction between the physical and digital realms (Lee, 2008; Möller, 2016). Through this integration, physical systems can be monitored, controlled, and optimised in real time. Petrolo, Loscri, and Mitton (2016) described CPS as a network of interconnected intelligent objects and systems capable of sensing, processing, and responding to environmental conditions. In the context of the built environment, CPS facilitates the coordination of building systems such as heating, ventilation, lighting, energy management, and environmental monitoring in smarter and more adaptive ways. This creates opportunities for improved efficiency, reliability, and responsiveness in building operations.

Within the built environment, CPS is particularly significant because buildings are inherently complex systems comprising numerous interdependent components. Mechanical, electrical, structural, environmental, and user-related systems interact continuously, and changes in one part of the building can affect performance in another. CPS offers a framework for better understanding and managing these interactions through data-driven integration and automation. Kan and Anumba (2019) argued that CPS represents an important stage in the digital transformation of construction and building management, especially as the sector evolves towards more connected and intelligent systems. In this context, CPS acts as a bridge between physical buildings and their digital representations, enabling smarter operations and more precise control of building performance.

Closely related to CPS, and often viewed as its more advanced manifestation, is DT technology. DT can be understood as a virtual representation of a physical asset, system, or process that is continuously updated through data exchanged from the physical counterpart (Glaessgen and Stargel, 2012; Grieves and Vickers, 2016; Tao and Qi, 2019). Rosen et al. (2015) noted that while the concept is rooted in earlier developments in simulation, 3D modelling, and prototyping, it has gained renewed significance because of advances in real-time sensing, computation, and system integration. Unlike static digital models, DT relies on continuous data flow to mirror the behaviour, state, and performance of physical systems. This allows for real-time monitoring, analysis, simulation, prediction, and optimisation.

The application of DT to the built environment requires digitally duplicating the physical asset into a virtual model that includes all relevant information about the building and its systems (Ito, 2019). This encompasses data relating to structure, services, energy, lighting, thermal systems, water use, occupancy, and indoor environmental conditions. Alonso et al. (2019) explained that a DT system for buildings aims to gather, process, and interpret critical lifecycle data so that decision-making is better informed and operational efficiency improved. The significance of DT in the built environment, therefore, lies in its ability to generate a continuously connected model of the building, thereby supporting more proactive and intelligent management.

Although the built environment is more complex and less predictable than controlled industrial systems, recent advances in digitalisation, sensor technology, and CPS have made the implementation of DT increasingly feasible (BSI Group, 2019; Khajavi et al., 2019; Opoku et al., 2021; Shahzad et al., 2022). Emerging technologies in the built environment have thus moved the sector beyond traditional digital tools towards integrated and intelligent systems that support real-time decision-making and lifecycle performance management. Among these technologies, CPS and DT stand out as important because they provide the conceptual and technical foundation for more responsive, data-driven, and sustainable building operations. This technological advancement is significant for indoor environments, where continuous monitoring, analysis, and optimisation are essential for enhancing building performance and occupant well-being.

1.5 Digital Twin for Indoor Environmental Quality Management

IEQ management has become increasingly important in the built environment because of the strong influence of indoor conditions on human health, comfort, well-being, and productivity. Since people spend most of their time indoors, the quality of indoor spaces directly affects daily life and organisational performance (Popkin, 1999; Klepeis et al., 2001; Cai et al., 2023). IEQ has therefore emerged as a key indicator of building performance, particularly for sustainable and user-centred buildings (Al-Sulaihi et al., 2015). However, maintaining acceptable indoor environmental conditions remains a complex task because IEQ is influenced by the interaction of several variables, including indoor air quality, thermal comfort, lighting, acoustics, building systems, climatic conditions, and occupant behaviour (Dykes and Baird, 2013; Altomonte et al., 2019; Devitofrancesco et al., 2019). This complexity calls for more advanced and integrated management approaches.

DT has emerged as a promising technology for addressing this challenge. Within the built environment, DT provides a digital representation of a physical building or facility that is continuously informed by real-time and historical data from the physical asset (Ito, 2019; Alonso et al., 2019). Unlike conventional digital models, DT is not static. It relies on data collected through sensors, monitoring systems, and interconnected digital infrastructure to reflect the changing condition and behaviour of the physical building (Tao and Qi, 2019). In this regard, DT offers a more dynamic and intelligent approach to building management, especially when the aim is to continuously monitor performance, anticipate potential problems, and optimise operational outcomes.

For IEQ management, DT's relevance lies in its ability to integrate, analyse, and respond to environmental data in real time. Data on temperature, humidity, air quality, lighting, ventilation, acoustics, energy consumption, and occupancy can be captured from the physical environment and transmitted to the virtual model for interpretation and simulation (Støre-Valen, 2019; Alonso et al., 2019). This enables building managers to obtain a more comprehensive understanding of indoor conditions and how they change over time. Habibi (2016) noted that digital innovation systems offer significant opportunities to improve IEQ, while Cai et al. (2023) highlighted the growing potential of DT for healthy indoor environments. Through continuous monitoring and predictive analysis, DT enables assessment of indoor environmental conditions not only as isolated measurements but also as part of a broader, evolving building performance system.

Another important strength of DT in IEQ management is its support for proactive rather than reactive decision-making. Traditional building management often depends on periodic inspections, occupant complaints, or delayed responses to performance issues. In contrast, DT enables building professionals to detect irregularities early, identify likely causes, and test possible interventions within the virtual environment before implementing them in the physical building. This can improve the speed and quality of decision-making while reducing unnecessary operational inefficiencies. Lolli et al. (2022) observed that the assessment and improvement of IEQ increasingly benefit from more integrated and evidence-based approaches, while Hosamo et al. (2023) demonstrated the relevance of DT frameworks for evaluating comfort performance and predicting faults in operational buildings.

DT also supports a two-way interaction between the physical and virtual environments. Changes occurring in the indoor environment can be mirrored in the virtual model, while recommendations or control actions generated by the model can inform building operations. This strengthens facility managers' ability to maintain indoor conditions at optimal levels and improve occupant comfort and well-being, while also supporting energy efficiency and operational performance (Cai et al., 2023). In essence, DT offers a more holistic and data-driven approach to IEQ management by integrating real-time sensing, virtual representation, analysis, and decision support into a single, integrated system. For this reason, DT is increasingly seen as a valuable technology for achieving healthier, smarter, and more responsive indoor environments.

1.6 Summary

This chapter laid the foundation for understanding how new-age technologies can support IEQ management in the built environment. It was established that the indoor environment is a critical aspect of human life because people spend most of their time indoors, and the quality of these spaces directly affects health, comfort, well-being, satisfaction, and productivity. The chapter also showed that IEQ is shaped by several interconnected factors, including air quality, thermal comfort, lighting, acoustics, and the influences of building design, operation, and occupant behaviour. The chapter further explained that effective management of indoor spaces depends on information-based decision-making. Buildings were presented as dynamic assets that require continuous monitoring, coordination, and informed intervention throughout their lifecycle. In this regard, the integration, communication, and use of building information were highlighted as essential for improving building performance and supporting user needs. The discussion then examined the growing role of digitalisation and emerging technologies in the built environment. It showed that technologies such as connected sensing systems, digital platforms, CPS, and DTs are transforming how buildings are managed. In particular, DT technology was identified as a major advancement for IEQ management because it enables continuous monitoring, analysis, and proactive decision-making. The chapter demonstrated that digital technologies provide a pathway towards healthier, smarter, and more responsive indoor environments.

References

Adeniran, A.A. and Shakantu, W. (2023) 'Digitalization of the South African construction industry: A potential for further future research', *Lecture Notes in Civil Engineering*, 245, pp. 17–27. doi:10.1007/978-3-030-97748-1_2

Agbo, K.E. et al. (2021) 'A review on ambient and indoor air pollution status in Africa', *Atmospheric Pollution Research*, 12(2), pp. 243–260. doi:10.1016/J.APR.2020.11.006

Aghimien, D., Aigbavboa, C. and Oke, A. (2018) 'Digitalisation for Effective Construction Project Delivery in South Africa', in *Contemporary Construction Conference: Dynamic and Innovative Built Environment*, pp. 3–10.

Alaloul, W.S. et al. (2020) 'Industrial Revolution 4.0 in the construction industry: Challenges and opportunities for stakeholders', *Ain Shams Engineering Journal*, 11(1), pp. 225–230. doi:10.1016/J.ASEJ.2019.08.010

Al-Hemoud, A. et al. (2018) 'Streamlining IAQ guidelines and investigating the effect of door opening/closing on concentrations of VOCs, formaldehyde, and NO2 in office buildings', *Building and Environment*, 137, pp. 127–137. doi:10.1016/J.BUILDENV.2018.03.029

Allen, J.G. et al. (2016) 'Associations of cognitive function scores with carbon dioxide, ventilation, and volatile organic compound exposures in office workers: A controlled exposure study of green and conventional office environments', *Environmental Health Perspectives*, 124(6), pp. 805–812. doi:10.1289/EHP.1510037

Alonso, R. et al. (2019) 'SPHERE: BIM digital twin platform', *Proceedings*, 20(9), pp. 1–5. doi:10.3390/proceedings2019020009

Al-Sulaihi, I. et al. (2015) 'Assessing indoor environmental quality of educational buildings using BIM', *Journal of Environmental Science and Engineering B*, 4(8). doi:10.17265/2162-5263/2015.08.006

Altomonte, S. et al. (2019) 'Indoor environmental quality and occupant satisfaction in green-certified buildings', *Building Research and Information*, 47(3), pp. 255–274. doi:10.1080/09613218.2018.1383715

Andargie, M.S., Touchie, M. and O'Brien, W. (2019) 'A review of factors affecting occupant comfort in multi-unit residential buildings', *Building and Environment*, 160, p. 106182. doi:10.1016/J.BUILDENV.2019.106182

Araszkiewicz, K. (2017) 'Digital technologies in facility management - The state of practice and research challenges', *Procedia Engineering*, pp. 1034–1042. doi:10.1016/j.proeng.2017.08.059

Aziz, N.D., Nawawi, A.H. and Ariff, N.R.M. (2016a) 'Building information modelling (BIM) in facilities management: Opportunities to be considered by facility managers', *Procedia - Social and Behavioral Sciences*, 234, pp. 353–362. doi:10.1016/j.sbspro.2016.10.252

Aziz, N.D., Nawawi, A.H. and Ariff, N.R.M. (2016b) 'ICT evolution in facilities management (FM): Building information modelling (BIM) as the latest technology', *Procedia - Social and Behavioral Sciences*, 234, pp. 363–371. doi:10.1016/j.sbspro.2016.10.253

Bakó-Biró, Z. et al. (2012) 'Ventilation rates in schools and pupils' performance', *Building and Environment*, 48(1), pp. 215–223. doi:10.1016/J.BUILDENV.2011.08.018

Botkina, D. et al. (2018) 'Digital twin of a cutting tool', in *51st CIRP Conference on Manufacturing Systems*, pp. 215–218. doi:10.1016/j.procir.2018.03.178

BSI Group (2019) Digital twins for the built environment. *BIMToday*. Available at: https://www.pbctoday.co.uk/news/bim-news/digital-twin-4-0/64519

Burge, P.S. (2004) 'Sick building syndrome', *Occupational and Environmental Medicine*, 61(2), pp. 185–190. doi:10.1136/OEM.2003.008813

Cai, J. et al. (2023) 'Digital twin for healthy indoor environment: A vision for the post-pandemic era', *Frontiers of Engineering Management* 2023, pp. 1–19. doi:10.1007/S42524-022-0244-Y

Devitofrancesco, A. et al. (2019) 'Development of an indoor environmental quality assessment tool for the rating of offices in real working conditions', *Sustainability*, 11(6), p. 1645. doi:10.3390/SU11061645

Dykes, C. and Baird, G. (2013) 'A review of questionnaire-based methods used for assessing and benchmarking indoor environmental quality', *Intelligent Buildings International*, 5(3), pp. 135–149. doi:10.1080/17508975.2013.783457

Fanger, P.O. (1970) *Thermal comfort. Analysis and applications in environmental engineering*. Copenhagen: Danish Technical Press.

Fennimore, J.P. (2014) *Sustainable facility management: operational strategies for today*. Available at: https://books.google.com/books/about/Sustainable_Facility_Management.html?id=GHTcMgEACAAJ (Accessed: 1 May 2023).

Frontczak, M. and Wargocki, P. (2011) 'Literature survey on how different factors influence human comfort in indoor environments', *Building and Environment*, 46(4), pp. 922–937. doi:10.1016/j.buildenv.2010.10.021

Galasiu, A.D. and Veitch, J.A. (2006) 'Occupant preferences and satisfaction with the luminous environment and control systems in daylit offices: A literature review', *Energy and Buildings*, 38(7), pp. 728–742. doi:10.1016/J.ENBUILD.2006.03.001

Garber, R. (2014) *BIM design: Realising the creative potential of building information modelling*. West Sussex, UK: Wiley. doi:10.1002/9781118914694

Garland, R.M. et al. (2021) 'The new WHO global air quality guidelines: What do they mean for South Africa?', *Clean Air Journal*, 31(2), pp. 1–2. doi:10.17159/CAJ/2020/31/2.12915

Glaessgen, E.H. and Stargel, D.S. (2012) 'The digital twin paradigm for future NASA and U.S. Air force vehicles', in *53rd Structures, Structural Dynamics, and Materials Conference*. American Institute of Aeronautics and Astronautics, pp. 1–14. doi:10.2514/6.2012-1818

Grieves, M. and Vickers, J. (2016) 'Digital twin: Mitigating unpredictable, undesirable emergent behavior in complex systems', in J. Kahlen, S. Flumerfelt and A. Alves (eds.), *Transdisciplinary perspectives on complex systems: New findings and approaches*. Cham, Switzerland: Springer, pp. 85–113. doi:10.1007/978-3-319-38756-7_4

Habibi, S. (2016) 'Smart innovation systems for indoor environmental quality (IEQ)', *Journal of Building Engineering*, 8, pp. 1–13. doi:10.1016/J.JOBE.2016.08.006

Haverinen-Shaughnessy, U. and Shaughnessy, R.J. (2015) 'Effects of classroom ventilation rate and temperature on students' test scores', *PLoS ONE*, 10(8). doi:10.1371/JOURNAL.PONE.0136165

Heinzerling, D. et al. (2013) 'Indoor environmental quality assessment models: A literature review and a proposed weighting and classification scheme', *Building and Environment*, 70, pp. 210–222. doi:10.1016/J.BUILDENV.2013.08.027

Hosamo, H.H. et al. (2023) 'Digital Twin framework for automated fault source detection and prediction for comfort performance evaluation of existing non-residential Norwegian buildings', *Energy and Buildings*, 281, p. 112732. doi:10.1016/J.ENBUILD.2022.112732

Hossain, M.A. and Nadeem, A. (2019) 'Towards digitizing the construction industry: State of the art of construction 4.0', in *Proceedings of the 10th International Structural Engineering and Construction Conference (ISEC)*. doi:10.14455/ISEC.RES.2019.184

Ismail, S.A. et al. (2022) 'Indoor air quality level influence sick building syndrome among occupants in educational buildings', *International Journal of Public Health Science (IJPHS)*, 11(2), pp. 503–517. doi:10.11591/IJPHS.V11I2.21125

ISO 7730 (2005) *Ergonomics of the thermal environment - analytical determination and interpretation of thermal comfort using calculation of the PMV and PPD indices and local thermal comfort criteria.* 3rd ed. Geneva: International Organisation for Standardization.

Ito, K. (2019) 'Change beyond building information modeling to a "Digital Twin" in architecture', *Japan Architectural Review*, 2(4), pp. 391–391. doi:10.1002/2475-8876.12123

Kan, C. and Anumba, C.J. (2019) 'Digital twins as the next phase of cyber-physical systems in construction', in *Proceedings of the ASCE International Conference on Computing in Civil Engineering*. Atlanta: American Society of Civil Engineers (ASCE), pp. 256–264. doi:10.1061/9780784482438.033

Khajavi, S.H. et al. (2019) 'Digital twin: Vision, benefits, boundaries, and creation for buildings', *IEEE Access*, 7. doi:10.1109/ACCESS.2019.2946515

Khovalyg, D. et al. (2020) 'Critical review of standards for indoor thermal environment and air quality', *Energy and Buildings*, 213, p. 109819. doi:10.1016/J.ENBUILD.2020.109819

Klepeis, N.E. et al. (2001) 'The National Human Activity Pattern Survey (NHAPS): A resource for assessing exposure to environmental pollutants', *Journal of Exposure Science & Environmental Epidemiology*, 11(3), pp. 231–252. doi:10.1038/sj.jea.7500165

Ko, C.-H., Pan, N.-F. and Chiou, C.-C. (2013) 'Web-based radio frequency identification facility management systems', *Structure and Infrastructure Engineering*, 9(5), pp. 465–480. doi:10.1080/15732479.2010.546804

Küller, R. et al. (2007) 'The impact of light and colour on psychological mood: A cross-cultural study of indoor work environments', *Ergonomics*, 49(14), pp. 1496–1507. doi:10.1080/00140130600858142

Kunz, J. and Fischer, M. (2020) 'Virtual design and construction', *Construction Management and Economics*, 38(4), pp. 355–363. doi:10.1080/01446193.2020.1714068

Lasi, H. et al. (2014) 'Industry 4.0', *Business and Information Systems Engineering*, 6(4), pp. 239–242.

Lee, E.A. (2008) 'Cyber physical systems: Design challenges', *Distributed Computing*, pp. 363–369. doi:10.1109/ISORC.2008.25

Lolli, F. et al. (2022) 'Post-occupancy evaluation's (POE) applications for improving indoor environment quality (IEQ)', *Toxics*, 10(10), p. 626. doi:10.3390/TOXICS10100626

Matandirotya, N.R. et al. (2022) 'State of ambient air quality in a low-income urban settlement of South Africa', *Scientific African*, 16, p. e01201. doi:10.1016/J.SCIAF.2022.E01201

Möller, D.P.F. (2016) *Introduction to cyber-physical systems*. Cham: Springer, pp. 81–139. doi:10.1007/978-3-319-25178-3_3

Nimlyat, P.S. (2018) 'Indoor environmental quality performance and occupants' satisfaction [IEQPOS] as assessment criteria for green healthcare building rating', *Building and Environment*, 144, pp. 598–610. doi:10.1016/J.BUILDENV.2018.09.003

Noor, M.N.M. and Pitt, M. (2009) 'A critical review on innovation in facilities management service delivery', *Facilities*, 27(5–6), pp. 211–228. doi:10.1108/02632770910944943

Opoku, D.G.J. et al. (2021) 'Digital twin application in the construction industry: A literature review', *Journal of Building Engineering*, 40, p. 102726. doi:10.1016/J.JOBE.2021.102726

Petrolo, R., Loscri, V. and Mitton, N. (2016) 'Cyber-physical objects as key elements for a smart cyber-city', in A. Guerrieri, V. Loscri, A. Rovella and G. Fortino (eds.), *Management of cyber physical objects in the future internet of things*. Cham, Switzerland: Springer, pp. 31–49. doi:10.1007/978-3-319-26869-9_2

Piasecki, M., Kostyrko, K.B. and Goljan, A. (2021) 'The ability to control VOC emissions from multilayer building materials', *Applied Sciences*, 11(11), p. 4806. doi:10.3390/APP11114806

Pishdad-Bozorgi, P. et al. (2018) 'Planning and developing facility management-enabled building information model (FM-enabled BIM)', *Automation in Construction*, 87, pp. 22–38. doi:10.1016/j.autcon.2017.12.004

Popkin, B.M. (1999) 'Urbanization, lifestyle changes and the nutrition transition', *World Development*, 27(11), pp. 1905–1916. doi:10.1016/S0305-750X(99)00094-7

Redlich, C.A., Sparer, J. and Cullen, M.R. (1997) 'Sick-building syndrome', *Lancet*, 349(9057), pp. 1013–1016. doi:10.1016/S0140-6736(96)07220-0

Rosen, R. et al. (2015) 'About the importance of autonomy and digital twins for the future of manufacturing', *IFAC-PapersOnLine*, 48(3), pp. 567–572. doi:10.1016/j.ifacol.2015.06.141

Sakhare, V.V. and Ralegaonkar, R.V. (2014) 'Indoor environmental quality: Review of parameters and assessment models', *Architectural Science Review*, pp. 147–154. doi:10.1080/00038628.2013.862609

Seppänen, O.A. and Fisk, W. (2006) 'Some quantitative relations between indoor environmental quality and work performance or health', *HVAC and R Research*, 12(4), pp. 957–973. doi:10.1080/10789669.2006.10391446

Shahzad, M. et al. (2022) 'Digital twins in built environments: An investigation of the characteristics, applications, and challenges', *Buildings*, 12(2), p. 120. doi:10.3390/BUILDINGS12020120

Stansfeld, S.A. and Matheson, M.P. (2003) 'Noise pollution: Non-auditory effects on health', *British Medical Bulletin*, 68, pp. 243–257. doi:10.1093/BMB/LDG033

Støre-Valen, M. (2019) 'Digitization of FM: Experiences from Norwegian property managers', in *European Facility Management Conference*, Dublin.

Tao, F. and Qi, Q. (2019) 'Make more digital twins', *Nature*. doi:10.1038/d41586-019-02849-1

Turunen, M. et al. (2014) 'Indoor environmental quality in school buildings, and the health and wellbeing of students', *International Journal of Hygiene and Environmental Health*, 217(7), pp. 733–739. doi:10.1016/j.ijheh.2014.03.002

Wei, W. et al. (2020) 'Review of parameters used to assess the quality of the indoor environment in Green Building certification schemes for offices and hotels', *Energy & Buildings*, 209, p. 109683. doi:10.1016/j.enbuild.2019.109683

Woodhead, R., Stephenson, P. and Morrey, D. (2018) 'Digital construction: From point solutions to IoT ecosystem', *Automation in Construction*, 93(March), pp. 35–46. doi:10.1016/j.autcon.2018.05.004

World Health Organization (2010) *WHO guidelines for indoor air quality: Selected pollutants*. Copenhagen, Denmark: World Health Organization. Available at: www.euro.who.int (Accessed: 6 February 2022).

2 Deep-Dive into Digital Twins

2.1 Introduction

Digital transformation has changed how industries understand, monitor, and manage physical systems. Within this transition, Digital Twin (DT) has emerged as one of the most significant technological advancements of the modern era. The concept has gained increasing attention because it offers a more sophisticated way to connect the physical and digital worlds through real-time data, simulation, analysis, and intelligent feedback. Although the idea of creating digital representations of physical systems is not entirely new, recent advances in sensing, data processing, modelling, and communication technologies have made DT more practical, dynamic, and broadly applicable across industries (Glaessgen and Stargel, 2012; Rosen et al., 2015; Tao and Qi, 2019). DT is therefore no longer seen merely as a modelling tool, but as an evolving technological framework for understanding and enhancing the performance of real-world systems.

The growing interest in DT is closely linked to the broader movement towards digitalisation and cyber-physical integration. As various industries continue to pursue greater efficiency, responsiveness, and predictive capabilities, there is increasing demand for technologies that provide a live, intelligent representation of physical assets and processes. DT addresses this need by creating a virtual counterpart that mirrors the condition, behaviour, and changes of a physical entity in near real time (Grieves and Vickers, 2016; Tao, Zhang and Nee, 2019d). This makes it possible to observe, simulate, predict, and optimise system performance in ways that traditional digital tools cannot easily match. Consequently, DT has been adopted in sectors such as aerospace, manufacturing, healthcare, transportation, energy, and the built environment (Tuegel et al., 2011; Kraft, 2016; Bhatti, Mohan and Raja Singh, 2021; Attaran and Celik, 2023).

A deep understanding of DT goes beyond a simple definition of the concept. It involves examining its origins, structure, operational logic, enabling technologies, and its relationship with related systems such as cyber-physical systems (CPS). DT is built on the capability to integrate physical assets, virtual models, data, and feedback mechanisms into a coherent system that can

DOI: 10.1201/9781003743330-2

continuously interact and adapt (Lee, 2008; Negri, Fumagalli and Macchi, 2017; Tao, Sui, et al., 2019b). It also relies on technologies like the Internet of Things (IoT), artificial intelligence (AI), cloud computing, extended reality (XR), and big data analytics to operate effectively in real-world environments (Kaur, Mishra and Maheshwari, 2020; Lu et al., 2020; Attaran and Celik, 2023). These technologies enable DT to evolve from static representation towards providing intelligent support for decision-making, optimisation, and innovation.

This chapter offers an in-depth exploration of DT by examining its core principles, practical framework, and technological enablers. It starts with digital transformation as the wider context in which DT has emerged. It then explores the concept and components of DT, including its process structure via the Deming Cycle. Subsequently, the chapter discusses CPS as a stepping stone to DT and highlights the key enabling technologies that support DT's development and operation. In doing so, it provides a comprehensive foundation for understanding DT as a transformative digital system that is increasingly relevant across various industries, including the built environment.

2.2 Digital Transformation

Hossain and Nadeem (2019) contend that the Fourth Industrial Revolution has reshaped industrial production through the application of digitalised processes. The skills required in the workplace have also been shaped by the history of industrial development, economic and social change, and technological advancement. As technology has advanced, the invention of innovative machinery and systems has increased industrial output and transformed the way work is performed (Hossain and Nadeem, 2019). The concepts of digitisation and digitalisation are rooted in this broader technological advancement.

Although they are often used together, the terms digitisation and digitalisation do not mean the same thing. Digitalisation remains a constantly evolving topic of discussion in industry, society, the economy, and academia. It is often confused with digitisation, even though the two concepts are distinct (Chapco-Wade-Safina, 2018). Savić (2019) also noted that the two terms are frequently used interchangeably because of their close resemblance. However, Chapco-Wade-Safina (2018) clearly distinguished between them, defining digitisation as the process of converting information from a physical format to a digital format, while digitalisation is the use of digitised information and digital technologies to improve business processes.

Savić (2019) further explained that digitisation primarily concerns data conversion. Its goal is to create digital formats by converting paper documents, photographs, films, microfilms, and Video Home System (VHS) tapes into digital files using encoding equipment and computers. Digitalisation, however, goes beyond simple conversion. In academia, it is often understood to include the wider effects of digital media and communication technologies on contemporary social life. In the industrial and organisational context, Bloomberg

(2018) described digitalisation as the transformation of an activity into a digital operation through the use of digital technologies. In this sense, digitalisation is not only about converting information but also about changing how activities are carried out.

Digitalisation creates opportunities for value generation, efficiency improvement, and new revenue streams. It places greater emphasis on information processing and the transformation of business operations through automation and digital systems (Savić, 2019). It is therefore concerned with new ways of doing business through the application of digitised information and information technology systems. Through this process, organisations can improve their value chains, productivity, efficiency, and even organisational culture. Digitalisation also affects all sectors of the economy as the adoption of emerging technologies increases. Different industries, occupations, and professions are therefore being reshaped by this transition.

The building industry is equally experiencing this digital revolution. As digital technologies continue to influence how projects are conceived, delivered, and managed, the skills required in the industry must also evolve to sustain efficiency and productivity. The shift is no longer limited to isolated software tools. It now includes broader digital processes that support coordination, automation, information exchange, and data-driven decision-making across the building life cycle. The digitalisation process is shown in Figure 2.1, which is adapted from Qi et al. (2019).

In the industrial sense, digitalisation has evolved through four major phases, as illustrated in Figure 2.1. The first phase is digital enablement, which involves the invention and use of computers for scientific computing and the conversion of physical documents into virtual formats. The second phase is digital assistance, which involves using computer-aided applications and systems to support tasks and improve efficiency. The third phase is digital interlinking and

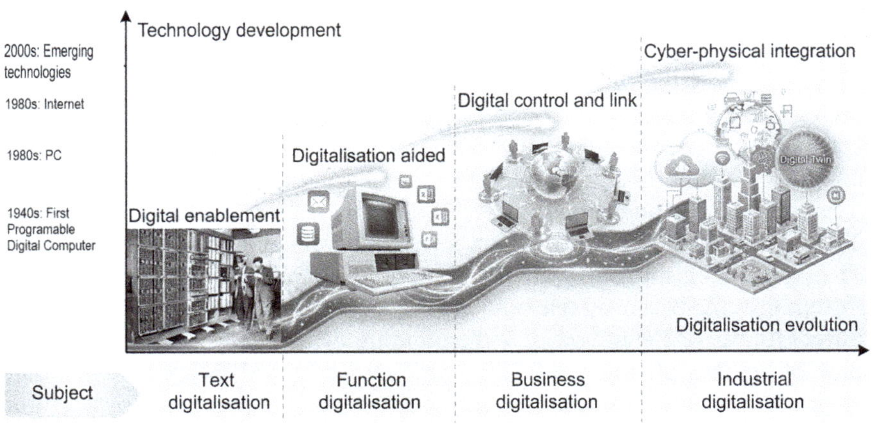

Figure 2.1 Process of digitalisation.

control, which uses the internet to enable faster and more efficient communication. The final phase is cyber-physical interlinkages, which incorporates new-generation technologies such as AI, the IoT, Big Data analytics, cloud computing, and DT to support and optimise activities at the industry level (Legner et al., 2017; Tao et al., 2019a).

Viewed in this way, digital transformation is not a single event but a progressive transition from basic digital conversion to intelligent, interconnected systems. It reflects the growing integration of digital technologies into industrial and organisational practice. This progression provides the foundation for more advanced technologies and more intelligent approaches to improve operational efficiency.

2.3 The Concept of Digital Twin

DT has been defined in various ways across the literature. It is important to recognise that the concept does not have a single fixed interpretation. Rather, it has been viewed from different disciplinary and industrial perspectives. According to the United States National Aeronautics and Space Administration's (NASA) 2010 Technology Roadmap, a DT is a predictive simulation of a replicated system that uses the best available history, sensor updates, and physical models to mirror the real-time activity of its corresponding twin (Shafto et al., 2010). This early description presents DT as a dynamic and data-informed representation of a physical system. By 2015, the concept had expanded beyond aerospace applications to encompass broader products and systems, leading to its wider adoption across industries such as informatics, engineering, and manufacturing (Negri, Fumagalli and Macchi, 2017).

Rosen et al. (2015) regarded DT as the next generation of modelling, simulation, and optimisation technologies, where the digital model is developed directly from the system itself. In this regard, DT is not merely a visual model. It is a platform for creating, developing, analysing, and visualising simulations. Some scholars have also described DT as a live model of a physical product together with its related descriptions and operational data. This means that DT can support the full lifecycle of a product or system, rather than merely serving as a digital reflection of its physical form (Glaessgen and Stargel, 2012; Kraft, 2016). The concept, therefore, extends beyond representation. It includes functionality, responsiveness, and lifecycle relevance.

Among the various definitions, the one provided by Glaessgen and Stargel (2012) remains the most widely accepted and frequently cited. They defined DT as a probabilistic multi-physics, multi-scale simulation of a complex product that uses the best available physical models, sensor updates, and other data sources to mirror the life of its corresponding twin. This definition highlights some of the key attributes of DT. These include simulation capability, data integration, real-time updating, and a close link between the physical and digital entities. A critical examination of the various definitions shows that DT is best understood as an emerging CPS. This is because it involves the bidirectional integration of computational and physical resources, in which embedded systems and communication networks enable virtual and physical components to connect and interact with one another (Lee, 2008). Grieves and Vickers (2016) explained that DT is built on three fundamental components: the physical entity, the virtual entity, and the data connections linking the two.

Dawkins, Hudson-Smith and Dennett (2018) described DT as a pairing of a physical entity with its digital replica, in which the digital replica perceives changes in the physical system and responds accordingly. This responsiveness occurs by generating data from the physical entity, which is then fed into the digital model to trigger a corresponding update. In the same way, information generated within the digital replica can also be sent back to the physical entity. This may occur automatically via actuators or manually via human intervention (Grieves and Vickers, 2016). In this sense, the idea of twinning reflects the continuous, interdependent interaction between physical and digital systems. DT can therefore be understood as a cyber-simulated representation of a physical object or system. Its purpose is to create a virtual clone that not only resembles the physical object but also behaves like it. At the same time, the DT can operate alongside the actual object in a coordinated and simultaneous manner (Jakob and Nilsson, 2018).

The concept of DT centres on creating a living digital counterpart to a physical system. It combines representation, simulation, data exchange, and feedback into a single integrated environment. This makes it more than a model and more than a simulation. It is an active digital system that can mirror, interpret, and influence the performance of its physical counterpart.

2.4 Cyber-Physical Systems as a Stepping Stone to Digital Twin

CPS play a significant role in the development of DT, acting as a stepping stone towards creating accurate and reliable virtual models of physical systems. CPS involve integrating physical systems with computing systems, allowing for the real-time monitoring, control, and analysis of physical processes (Ning and Liu, 2015; Kan and Anumba, 2019). CPS is a foundational technology that enables DT by providing a real-time view of physical systems. This view can be used to create an accurate digital replica of the physical system, which can be analysed and optimised for efficiency, safety, and performance (Negri, Fumagalli and Macchi, 2017; Kan and Anumba, 2019).

2.4.1 Understanding the Concept of CPS

The advent of the Fourth Industrial Revolution (4IR) has introduced a wide range of data exchange systems and interconnected sensory networks for industrial use (Leitão et al., 2015). This development has given rise to a cluster of intelligent, interconnected technologies collectively referred to as CPS, which are increasingly embedded in 21st-century industries (Vogel-Heuser et al., 2017). CPS can be understood as systems designed to monitor ongoing situations, interpret operating conditions, and take intelligent actions based on that evaluation. This ability depends on a sophisticated chain of sensors, processors, communication links, and computational models that work together within one operational environment (Thoben et al., 2017).

The emergence of CPS has created new possibilities for addressing complex industrial problems through timely, intelligent responses to operational changes or disruptions (Lee and Seshia, 2017). At its core, CPS links the physical and digital worlds, enabling the condition and behaviour of physical assets to be continuously sensed, communicated, interpreted, and acted upon. In practical terms, this means that physical systems such as machines, equipment, or production lines can be observed through sensors, while their digital counterparts receive and process the resulting information. This close interaction between physical and digital systems enables smarter operation, more informed control, and improved system responsiveness. It also creates the basis for real-time replication of physical systems in digital form. This real-time replication is commonly associated with DT, which may be seen as an advanced expression of cyber-physical interaction (Lee and Seshia, 2017). In this regard, the study by Ait-Alla et al. (2019) is important because it examined the relationship between the physical and virtual worlds within the cyber-physical spectrum and used simulation models to explore how systems perform under different conditions.

Within the literature, many definitions have been proposed for the digitalisation of products, workpieces, machines, complete production lines, and industrial systems. In recent years, related and sometimes overlapping terms such as DT (Lee, Bagheri and Kao, 2015), digital shadows, and digital avatars (Hribernik et al., 2013) have also emerged within the broader discourse surrounding CPS. Although these concepts differ in their levels of sophistication and functionality, they all point to the growing effort to create stronger, more intelligent links between physical systems and digital representations. The concept of CPS itself can be traced to Lee (2008), while its structure was more comprehensively outlined by Lee, Bagheri and Kao (2015). Their structure presents CPS as a five-stage architecture in which each stage builds on the capabilities established in the preceding one.

The CPS architecture progresses through five connected stages. It begins with Smart Connection, where relevant data are collected from the physical environment through sensors and linked systems. These data are then transformed into useful information in the Data-to-Information Conversion stage for monitoring, diagnosis, and forecasting. At the Cyber layer, information is stored and organised into digital models and historical records. The Cognition stage enables intelligent analysis, simulation, and decision-making. Finally, Configuration coordinates multiple CPS into a unified, self-optimising system that supports autonomous control and operational efficiency (Lee, Bagheri and Kao, 2015).

The significance of CPS lies in its ability to integrate sensing, communication, modelling, analysis, and control within a single technological framework.

CPS does not merely collect information from the physical environment. It also transforms that information into models, supports intelligent decision-making, and coordinates multiple systems into a single integrated operational environment. This is what makes CPS such an important technological development in modern industry. It improves the intelligence, responsiveness, and efficiency of systems while also laying the groundwork for more advanced applications such as DT. In this sense, CPS can be regarded as both a foundational concept and a practical framework for the digital transformation of industrial systems.

2.4.2 CPS Contributions to Digital Twin

CPS serves as an important stepping stone towards DT by enabling the creation of an accurate and responsive digital representation of a physical system. Through CPS, physical systems can be observed, connected, analysed, and controlled in ways that support the development of a reliable DT. This makes CPS a foundational layer upon which DT can be built. Once established, the DT can then be used for simulation, analysis, optimisation, forecasting, and training. By using CPS as a pathway to DT, engineers and system managers can reduce risk, improve reliability, and optimise system performance (Ning and Liu, 2015; Kan and Anumba, 2019). CPS contributes to DT in several important ways. One of the most significant is real-time monitoring. CPS enables physical systems to be continuously monitored through sensors and other connected devices. The data collected from these sources can then be used to build an accurate, up-to-date digital representation of the physical system. This real-time data stream is essential for the creation and operation of a meaningful DT (Negri, Fumagalli and Macchi, 2017).

Another major contribution is predictive maintenance. CPS enables the detection of patterns in system performance and the identification of when maintenance is likely to be required. This supports a proactive approach to maintenance rather than a reactive one. As a result, downtime can be reduced, system reliability can be improved, and the condition of the physical system can be reflected more accurately in the DT (Wang et al., 2020). CPS also supports performance optimisation. Because it provides continuous data and system visibility, engineers can simulate different scenarios, test alternatives, and assess the likely impact of possible changes. By making adjustments in the DT, they can predict how those changes may affect the physical system before implementation. This supports more informed and effective decision-making (Negri, Fumagalli and Macchi, 2017). In addition, CPS enhances safety. It can detect anomalies, unexpected behaviour, or abnormal operating conditions within physical systems and alert operators before major incidents occur. This allows preventive action to be taken early, thereby reducing the likelihood of accidents and minimising the consequences of disruptions when they arise (Zhou et al., 2019; Kan and Anumba, 2019).

CPS further contributes to improved efficiency. Through the continuous monitoring of operations, including energy use and system behaviour, CPS can help identify areas where resources are being wasted or where performance can be improved. This can lead to cost savings, reduced environmental impact, and better use of system resources (Tao et al., 2019d; Kan and Anumba, 2019). CPS is a foundational technology that plays a critical role in enabling DT. By supplying real-time data from physical systems, CPS allows the creation of accurate digital replicas that can support simulation, analysis, optimisation, and training. Its contributions to monitoring, maintenance, safety, efficiency, and performance make it an essential technology for the successful development and operation of DT across industries.

2.5 Composition of Digital Twin

Studies have observed that DT is a dynamic cyber-physical technology (Tao, Sui, et al., 2019b). At a minimum, DT requires the capabilities outlined in the fourth stage of the CPS architecture, as highlighted by Lee, Bagheri, and Kao (2015), which is tagged the cognition stage. These capabilities are essential for establishing intelligent control of industrial processes. As shown in Figure 2.2, adapted from Ait-Alla et al. (2019), traditional systems are limited to the physical process itself. In such systems, data generated from the physical process is often stored only in databases or other external storage systems, without deeper integration into live virtual environments (Ait-Alla et al., 2019). Recent advances in CPS and other Fourth Industrial Revolution concepts now allow physical and digital pieces to coexist. In this setup, data is transmitted continuously from the physical entity to its virtual replica. When a DT receives current information, it can mimic, support, or even predict decisions within the existing system. This creates greater opportunities for decentralised control systems in future industrial applications (Scholz-Reiter and Freitag, 2007).

DT refers to a physical asset and a digitised asset model that can communicate through forward and backward interactions and progress together in

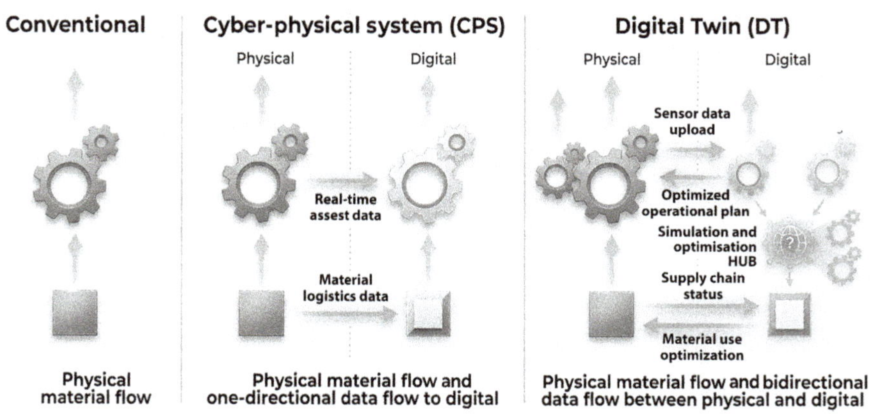

Figure 2.2 Progression to DT.

unison (Tao, Sui, et al., 2019b). Advances in digitisation technologies have enabled the collection and storage of data from physical assets in a more holistic manner. This ensures that the virtual model conforms accurately to the physical asset in real time (Tuegel et al., 2011). In essence, the strength of DT lies in its ability to connect the real and virtual worlds in a synchronised and functional manner. The value of DT is immense because it enables the dynamic replication of physical assets in virtual form. This replication can reflect the physical structure, position, state, arrangement, condition, attitude, and motion of the asset (Qi et al., 2019). A true digital replication of the physical world can be achieved through sensor-based data collection, 3D computing, Big Data analytics, AI, and machine learning. These capabilities allow DT to support a wide range of functions, including system observation, diagnosis, prediction, and optimisation (Cai et al., 2017). In practical terms, DT provides a richer and more intelligent representation of a system than conventional digital models.

DT can also support decision-making across many industries by assessing current conditions, analysing past challenges, and projecting future uncertainties (Tao et al., 2019a). In addition, the technology can support the orientation, training, and development of both operational and administrative workers. This is possible because the virtual model is a live and realistic representation of the physical structure, attributes, and behaviour of a real-world asset (Botkina et al., 2018). DT also helps users to understand complex systems and processes more clearly. Simulation and virtual reality (VR) tools can break down physical assets and procedures into smaller, more manageable parts (Qi et al., 2019). A further benefit is that DT can capture and digitise the knowledge, skills, and experience of industry experts. This knowledge can be stored, shared, and updated over time, thereby helping to reduce knowledge gaps in a given field (Qi et al., 2019).

The composition of DT is often explained in terms of five core aspects: physical entities, virtual models, twin data, services, and connections. Physical entities are the real-world assets for which virtual models are created. Virtual models are real-time digital replicas of physical entities (Tao, Sui, et al., 2019b). Twin data are the driving force behind the entire technology. These data are often multi-dimensional and heterogeneous in nature. Some of it is collected directly from physical assets, while some is generated by digital models through simulation. Data may also come from system processes and expert knowledge (Qi et al., 2018). DT services include the various functions performed through the technology, such as observation, diagnosis, forecasting, management, and optimisation. Lastly, DT connections refer to the links between the physical asset, the virtual replica, the data, and the services. These connections enable effective operation, complex analysis, and multi-layered simulation (Qi et al., 2019; Tao et al., 2019b).

DT has emerged as a modern expression of CPS because it unifies elements from physical and virtual space and creates cross-functional links between them. This twinning process is largely enabled by sensor data, which links the real and virtual worlds (Negri, Fumagalli and Macchi, 2017). For a DT to be both geometrically and functionally accurate, it must incorporate all relevant geometric and functional data of the physical object (Grieves and Vickers, 2016). Sensor data is used to create a graphical and virtual representation of the physical product. This representation can then be used for modelling, visual inspection, analytical evaluation, simulation, and future planning (Zhu, Liu and Xu, 2019). One of the most significant characteristics of DT is its ability to adapt dynamically in near real time as the condition of the physical product changes (Tao, Zhang and Nee, 2019h). Sensor data can also be used to show how the physical product is influenced by its environment and, in turn, how it affects that environment (Grieves and Vickers, 2016). Tao et al. (2018) further identified real-time reflection as a core characteristic of DT. With the aid of DT, real-world data can be replicated in the virtual world with high accuracy and synchronisation (Talkhestani et al., 2018).

Another important feature of DT is its ability to self-evolve. The virtual model changes dynamically through constant comparison between the physical and digital environments, ensuring that relevant attributes and components are accurately replicated (Tuegel et al., 2011). DT also strengthens the connection between these two spaces. Gabor et al. (2016) described this attribute as 'convergence and interaction'. Tao et al. (2018) extended this explanation in two ways. First, convergence and interaction occur through the integration of physical space data collected at different times. Second, historical data are continuously integrated with current data. Through this continuous exchange, DT improves its capacity to self-evolve and remain relevant over time.

Figure 2.3 presents a typical DT framework that can be adapted to different industries. The framework illustrates the interaction between the physical and digital worlds. It also highlights the parameters that should be documented within the system. In addition, it shows the performance measures that DT can

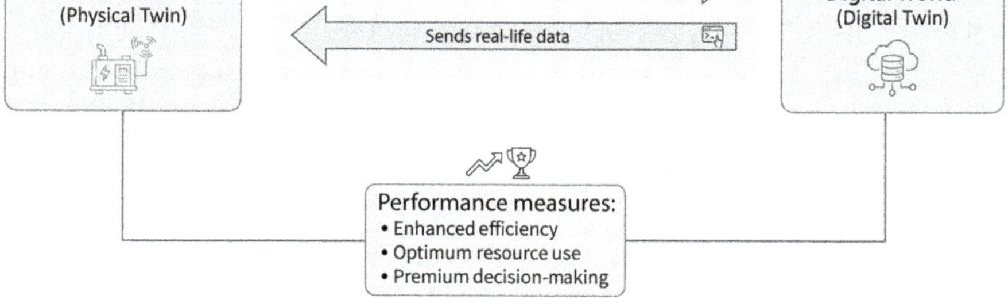

Figure 2.3 Typical Digital Twin framework.

help to improve. DT has enabled new technologies to influence decision-making through advanced and dynamic modelling. This is achieved through its ability to provide a comprehensive visualisation of the entire system (Dawkins, Hudson-Smith and Dennett, 2018). A variety of industries have already adopted DT. The aerospace industry was among the first to adopt the concept, which then expanded to manufacturing, product lifecycle management, industrial engineering, robotics, and informatics (Kan and Anumba, 2019).

2.5.1 Basic Requirements for Deploying Digital Twin

A Digital Twin System (DTS) is a dynamic virtual replica of a physical object, asset, or process. It is designed to simulate the behaviour and performance of its real-world counterpart. The purpose of a DTS is not only to mirror reality but also to optimise system performance, predict failures, and generate new insights for improved decision-making. To achieve these outcomes, a DT must satisfy several basic requirements. These include autonomy, connectivity, real-time operation, reliability, an interoperable platform, data exchange and integration, resource optimisation, adaptability, and data security.

1 **Autonomy**: A robust DTS must be capable of making autonomous decisions based on the real-time data it receives. This allows the system to operate more efficiently and effectively, while reducing dependence on constant human intervention and limiting the risk of avoidable errors. Autonomy is achieved through AI and machine learning algorithms, which enable the DTS to learn from incoming data and adapt to changing operating conditions (Rosen et al., 2015).
2 **Connectivity**: Connectivity is essential for a DTS because it must gather data from multiple sources, including sensors, actuators, and other connected devices. It must also communicate with other systems where necessary. To support this, a strong and reliable network infrastructure is required. Such infrastructure ensures seamless data transmission and keeps the DTS continuously updated with the latest information from its physical counterpart (Stark, Fresemann and Lindow, 2019; Singh et al., 2022).
3 **Real-time operation**: Real-time operation is a critical requirement because it enables the DTS to provide timely and accurate insight into the condition and performance of the physical system. Through continuous monitoring and analysis of live data, the DTS can detect problems early, respond quickly, and support performance optimisation while reducing downtime (Uhlemann et al., 2017b; Uhlemann, Lehmann and Steinhilper, 2017a; Samir, Maffei and Onori, 2019; Tao et al., 2019a).
4 **Reliability**: A DTS must be reliable if it is to provide trustworthy and consistent information. Reliability is achieved through careful validation and verification processes that ensure alignment between the DT and the real-world system. In addition, the DTS should be able to manage faults and failures without collapsing, so that the system remains functional and

dependable during operation (Khajavi et al., 2019; Samir, Maffei and Onori, 2019; Tao, Zhang and Nee, 2019g).

5 **Interoperable platform**: An interoperable platform is necessary to ensure the smooth integration of different hardware, software applications, and communication protocols. This enables the DTS to interact with a wide range of systems and devices. It also supports seamless data exchange and strengthens the effectiveness of the DT environment (Malakuti et al., 2018; Tao, Zhang and Nee, 2019g; Sun et al., 2020).
6 **Data exchange and integration**: A DTS must be able to exchange, combine, process, and interpret data from multiple sources. This is essential for developing a complete understanding of system performance. Because DT environments often involve large volumes of heterogeneous data, the system must be capable of handling diverse data formats while ensuring that all relevant information is incorporated into analysis and decision-making (Fortino et al., 2016; Kaewunruen and Lian, 2019).
7 **Resource optimisation**: One of the major objectives of a DTS is to optimise the use of available resources, including energy, materials, time, and personnel. This is achieved through continuous monitoring, performance analysis, and identification of areas for efficiency improvement. By doing so, the DTS helps to reduce waste, improve productivity, and enhance operational performance (Uhlemann, Lehmann and Steinhilper, 2017a; LaGrange, 2019; Tao and Qi, 2019).
8 **Adaptability**: A DTS must be adaptable if it is to remain useful under changing conditions. Systems, environments, and operational demands do not remain constant, and the DT must be able to respond accordingly. Machine learning and AI technologies support this requirement by enabling the DTS to learn from new data, refine its internal models, and improve its predictive and decision-making abilities over time (Vathoopan et al., 2018; Woodhead, Stephenson and Morrey, 2018).
9 **Data security**: Because a DTS collects, stores, and processes sensitive data relating to physical systems and operations, data security is essential. The system must use strong encryption, authentication, and access-control measures to prevent unauthorised access and to preserve data privacy, integrity, and system trustworthiness (Ning and Liu, 2015; Shahzad et al., 2022).

A successful DTS must satisfy these core requirements to function effectively and reliably. When these requirements are properly addressed, the system can provide valuable insight into performance, support operational optimisation, reduce cost, and improve decision-making. As DT continues to advance across industries, understanding and implementing these requirements becomes increasingly important for successful deployment.

2.5.2 Digital Twin Process Structure Based on the Deming Cycle

A DT is a cyber-simulated representation of a physical object or system. It exists to create a virtual clone of an existing object that looks, behaves, and responds like its physical counterpart. At the same time, the DT can operate alongside the actual object in a coordinated and simultaneous manner. Jakob and Nilsson (2018) suggested that DT can be understood through an adaptation of Deming's theory of management, commonly known as the Deming Cycle. Lontsikh et al. (2020) also adapted Deming's management theory to develop digital management approaches for automation processes. In this regard, the Deming Cycle provides a useful structure for understanding how DT processes can be organised, improved, and sustained. The Deming Cycle is illustrated in Figure 2.4.

The Deming Cycle, also known as the Plan-Do-Check-Act (PDCA) cycle, is a continuous improvement method that emphasises iterative, incremental improvements to systems and processes. It is widely used in management and operational contexts because it provides a structured approach to identifying problems, implementing solutions, evaluating results, and standardising improvements. In the context of DT, the Deming Cycle offers a practical framework for streamlining the implementation, operation, and optimisation of the DT process (Lontsikh et al., 2020). This is particularly important because DT is not a static system. It evolves continuously through data exchange, analysis, feedback, and system refinement.

Applying the Deming Cycle to DT helps explain how the DT can be developed and managed systematically. Each stage of the cycle contributes to the overall success of the system by supporting planning, execution, evaluation, and continuous improvement. Through this lens, DT becomes more than a digital replica. It becomes an adaptive management system capable of learning from performance, responding to change, and improving over time. The Deming management theory can therefore be explored in relation to DT by

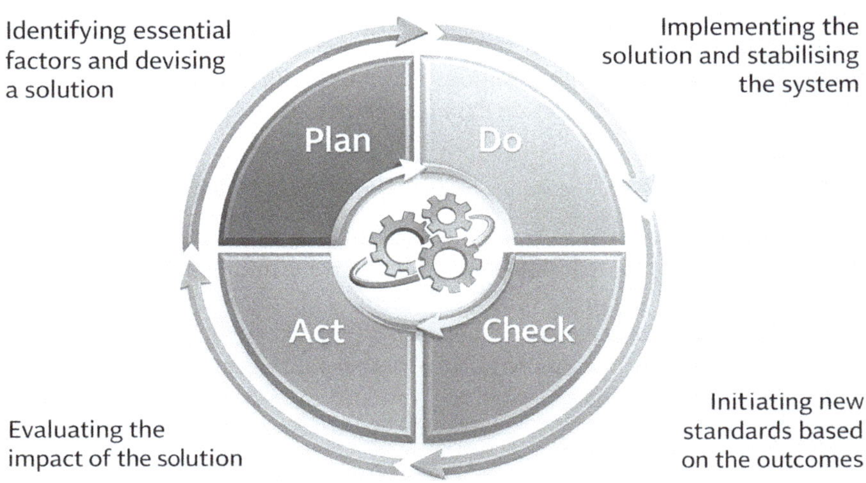

Figure 2.4 The Deming cycle.

examining how each phase of the cycle contributes to the effectiveness, reliability, and continuous development of the DT process.

2.5.2.1 Plan: Identifying Essential Factors and Devising a Solution

The planning phase involves identifying the key factors required for the successful implementation of a DT. This includes defining the DT's objectives, selecting the most suitable technologies and data sources, and determining the required level of fidelity for the digital representation. This phase also involves devising a practical solution for integrating the DT with the physical system and outlining the steps necessary to achieve the intended outcomes. In this sense, the planning phase provides the foundation upon which the entire DTS is built. If planning is weak, the system's deployment and operation are likely to be ineffective (Jakob and Nilsson, 2018). For a DTS, the planning stage may involve defining the scope and goals of the DT project, identifying relevant stakeholders and understanding their needs, selecting suitable technology platforms and data sources, and establishing clear performance metrics and targets to guide the system's development and evaluation.

2.5.2.2 Do: Implementing the Solution and Stabilising the System

Once the planning phase is complete, the next step is to implement the proposed solution. This phase involves deploying the DT, integrating it with the physical system, and stabilising it to ensure consistent, reliable performance. At this stage, data collection and monitoring begin to gather sufficient information for analysis, control, and future improvement. The emphasis of this phase is on execution, system coordination, and operational stability (Jakob and Nilsson, 2018). For a DTS, the Do phase may involve developing the DT model and integrating it with real-world data sources, implementing necessary communication protocols and data-processing systems, establishing control mechanisms that support autonomous decision-making, and monitoring system performance while collecting data for further analysis.

2.5.2.3 Check: Evaluating the Impact of the Solution

The Check phase involves assessing the DT's performance to determine whether the implemented solution has effectively addressed the identified issues or produced meaningful improvement. This phase requires analysing the collected data, comparing actual results against predefined metrics and targets, and identifying areas requiring further refinement. In this stage, the focus shifts from implementation to evaluation. The purpose is to understand what is working, what is not working, and why (Jakob and Nilsson, 2018). For a DTS, the Check phase may involve analysing the data collected during the Do phase, comparing the DT's performance against predefined metrics and targets, identifying areas where the system has performed well or underperformed, and determining the root causes of any discrepancies or issues that may have emerged.

2.5.2.4 Act: Initiating New Standards Based on the Outcomes

The Act phase is the final stage of the Deming Cycle. It involves using insights gained from the Check phase to refine the DT process, implement improvements, and establish new operating standards informed by observed outcomes. These improvements are then incorporated into the existing system, after which the cycle begins again. In this way, the DT process remains dynamic and continuously improves over time. This phase ensures that lessons learned are not lost but are translated into practical changes that strengthen future performance (Jakob and Nilsson, 2018). For a DTS, the Act phase may involve updating the DT model based on the insights gained from the Check phase, implementing improvements to the DT process through refined algorithms or enhanced data integration, establishing new standards or revised operational procedures based on the outcomes of the earlier phases, and repeating the cycle to support continuous improvement and adaptation over time.

All put together, the Deming Cycle provides a systematic framework for continuous improvement in DT processes. By repeatedly planning, implementing, evaluating, and acting on improvements, organisations can optimise the performance of their DT and strengthen their capacity to make informed decisions, predict outcomes, and use resources more effectively (Lontsikh et al., 2020). Integrating the Deming Cycle into the DT process helps to ensure that the system remains relevant, adaptive, and effective over time.

2.6 Enabling Technologies for Digital Twin

DT is a powerful technology transforming how industries operate by creating virtual replicas of physical objects, processes, or systems. It enables the visualisation, analysis, and optimisation of different scenarios before physical implementation. This can save time and resources while improving efficiency, productivity, and decision-making. DT is increasingly being applied across several industries, including manufacturing, healthcare, transportation, aerospace, energy, and the built environment (Malakuti et al., 2018; Lu et al., 2020; Singh et al., 2022; Attaran and Celik, 2023). Its growing relevance lies in its ability to connect physical and digital systems in ways that support monitoring, simulation, prediction, and operational improvement (Grieves and Vickers, 2016; Tao and Qi, 2019; Shahzad et al., 2022).

To make this possible, several enabling technologies have emerged and continue to evolve in support of DT development and deployment. These technologies form the backbone of DT because they provide the infrastructure, connectivity, intelligence, and computing capacity required to efficiently build, update, and operate DT models (Qi et al., 2019; Kaur, Mishra, and Maheshwari, 2020; Attaran and Celik, 2023). Without them, DT would remain a static digital model rather than a dynamic and responsive system. In practice, these technologies enable the DT to gather data from the physical world, process and interpret that data, simulate possible outcomes, and support timely decision-making (Lu et al., 2020; Yu et al., 2022).

The key enabling technologies for DT include the IoT, big data analytics, AI, cloud computing, and XR (Attaran and Celik, 2023). IoT provides the sensing and communication layer that links physical assets to digital environments and supports the continuous flow of real-time data (Kaur, Mishra and Maheshwari, 2020; Qi et al., 2019). Big data analytics enables the handling, organisation, and interpretation of the large volumes of structured and unstructured data generated by DT environments (Tao, Zhang and Nee, 2019f; Wang et al., 2020; Yu et al., 2022). AI strengthens DT by supporting learning, pattern recognition, prediction, reasoning, and intelligent decision support (Min et al., 2019; Kaur, Mishra and Maheshwari, 2020; Attaran and Celik, 2023). Cloud computing provides scalable storage, computational power, remote accessibility, and collaborative capacity, all of which are necessary for managing complex DTS (Alam and El Saddik, 2017; Hu et al., 2018; Tao, Zhang and Nee, 2019g). XR, including VR, augmented reality (AR), and related immersive tools, improves users' interaction with DT by enabling enhanced visualisation, simulation, interpretation, training, and operational engagement (Rabah et al., 2018; Tao, Zhang and Nee, 2019i; Zhu, Liu and Xu, 2019).

The strength of DT does not come from a single technology, but from integrating these enabling technologies into a coordinated digital ecosystem. Their combined contribution allows DT to function as an intelligent, connected, adaptive, and continuously evolving system rather than a simple digital representation (Qi et al., 2018; Lu et al., 2020; Wanasinghe et al., 2020). As these technologies continue to mature, DT is expected to play an even greater role in helping organisations optimise operations, improve system performance, reduce uncertainty, and enhance efficiency and productivity across different sectors (Tao and Qi, 2019; Sleiti, Kapat and Vesely, 2022; Attaran and Celik, 2023). In this sense, enabling technologies are not merely supportive tools for DT, but the essential foundation upon which effective DT development and deployment depend.

2.6.1 Internet of Things

The IoT refers to a network of physical devices, sensors, and other objects connected to the Internet and able to communicate with one another. The data generated by IoT devices provides real-time information about the performance, behaviour, and condition of physical objects. This allows organisations to optimise their operations and make more informed decisions (Atlam and Wills, 2020). IoT is one of the most important enabling technologies for DT because it provides the data needed to create, update, and implement DT models. IoT devices can generate vast amounts of data, which can be used to create DT models of physical objects and systems. These models can then be used to monitor and analyse the performance of physical objects and systems, identify potential issues, and optimise operations. IoT also enables the integration of different data sources, including sensors, control systems, and other connected devices, thereby creating a more comprehensive view of the physical system (Mateev, 2020).

One of the key benefits of IoT as an enabling technology for DT is its ability to provide real-time data. IoT devices can deliver continuous updates on the performance and behaviour of physical objects, enabling DT models to reflect the system's current condition with high accuracy. Real-time data also allows organisations to identify issues quickly and respond to them promptly. This improves operational efficiency and reduces downtime (Tao, Zhang, et al., 2019a). IoT also supports predictive maintenance by signalling when maintenance may be required based on real-time operational data. This allows organisations to move from reactive maintenance to proactive maintenance. Predictive maintenance can help optimise maintenance schedules, reduce downtime, and extend the lifespan of physical objects and systems. In this regard, DT models can be used to simulate different maintenance scenarios and identify the most effective course of action (Wang et al., 2020).

Another major benefit of IoT as an enabling technology for DT is its ability to support remote monitoring and control. IoT devices can monitor and control physical objects and systems from virtually any location, enabling remote teams to collaborate and make informed decisions based on live data. This is particularly beneficial for organisations that manage complex, large-scale, or geographically distributed systems (Tao, Zhang and Nee, 2019e). In addition, IoT can support the creation of DT models for consumer products. Smart home devices, wearables, and other connected consumer products can generate data that can be used to create DT models for those products. These models can then be used to optimise product design, improve user experience, and identify potential problems before they occur (Kaur et al., 2020).

IoT is therefore an essential enabling technology for DT. IoT devices provide the real-time and continuous data needed to create DT models of physical objects and systems. These models help organisations optimise operations, reduce downtime, improve maintenance planning, and extend the useful life of assets and systems. IoT also supports remote monitoring and control, predictive maintenance, and the development of DT models for consumer products.

2.6.2 Artificial Intelligence

AI has been identified as one of the key enablers of DT because of its ability to analyse large volumes of data and generate insights that can be used to optimise operations and improve system performance (Attaran and Celik, 2023). AI involves the use of algorithms, models, and computational techniques to process data, identify patterns, learn from observations, and support intelligent decision-making. In the context of DT, AI can be used to analyse the data generated by sensors and other IoT devices and convert that data into meaningful information that supports optimisation, prediction, and control (Mateev, 2020). This makes AI an essential layer of intelligence within the DT environment.

AI enables DT by providing the analytical capability needed to create, update, and maintain the digital replica of a physical asset or system. Because DT environments often involve large, complex, and continuously changing datasets, AI is important for managing this complexity and extracting useful knowledge. AI also supports the development of more accurate and detailed DT models by efficiently processing large amounts of information and identifying patterns that may not be easily recognised through conventional analysis (Fan et al., 2019). In this way, AI strengthens the predictive, adaptive, and decision-support capabilities of DT.

AI-enabled DT can be applied across different industries, including manufacturing, energy, transportation, and healthcare. In manufacturing, for example, AI-enabled DT can be used to monitor machine condition and performance, predict when maintenance will be required, and optimise production processes. In energy and transportation systems, AI can help improve asset performance, support operational planning, and enhance response to changing conditions. In healthcare, AI-enabled DT can be used to create a virtual representation of a patient, which can then support health monitoring, diagnosis, and the simulation of different treatment scenarios (Attaran and Celik, 2023). These examples show that AI expands the practical value of DT by making it more responsive and intelligent in different contexts.

One of the most important benefits of AI-enabled DT is its support for predictive maintenance. By analysing data generated by sensors and other IoT devices, AI models can predict when maintenance is likely to be required and recommend the most suitable maintenance schedule. This helps organisations reduce downtime, extend the lifespan of physical assets, and optimise maintenance costs (Tao et al., 2019a). AI, therefore, allows DT to move beyond observation and into anticipation, which is one of its most valuable operational strengths. Another major benefit of AI-enabled DT is its support for real-time decision-making. By continuously analysing data in real time, AI models can generate insights that support timely, informed decisions about operations, performance, and system behaviour. This can be especially beneficial for organisations that manage distributed or complex assets, such as those in the transportation and energy sectors (Xu et al., 2018). In such cases, AI allows DT to provide intelligent support under dynamic operating conditions.

AI is therefore critical to the effective functioning of DT. It provides the analytical intelligence required to create and maintain meaningful digital replicas of physical assets and systems. AI-enabled DT can support predictive maintenance, real-time decision-making, and improved operational performance across different industries. As AI technology advances, its importance as an enabling technology for DT will grow.

2.6.3 Extended Reality

The use of AR and VR has been identified as a key enabler of DT, as it enhances the visualisation of virtual models and provides a more immersive user experience (Zhu, Liu and Xu, 2019; Attaran and Celik, 2023). XR is a broader term that encompasses immersive digital technologies, including AR, VR, and related interactive environments. XR technology uses digital interfaces either to enhance the user's perception of the real world or to create a fully immersive virtual environment. In the context of DT, AR/VR can be used to improve how the virtual model is viewed, understood, and experienced by the user (Dawkins et al., 2018).

XR plays a major role in enabling DT by providing a more realistic, interactive, and immersive representation of the virtual model. Through AR/VR, the virtual model can be visualised in real time, allowing users to interact with it more naturally and intuitively (Tao et al., 2019c). This improves understanding of the system and makes it easier to interpret complex information. XR-enabled DT can be applied in various industries, including manufacturing, construction, and healthcare. In manufacturing, AR/VR can be used to visualise the virtual model of a machine or an assembly line, allowing the user to engage with it more directly and intuitively. In construction, AR/VR can be used to visualise the virtual model of a building or infrastructure project, allowing users to explore design options and better understand spatial relationships before physical implementation (Attaran and Celik, 2023; Zhu et al., 2019).

One of the key benefits of XR-enabled DT is its ability to support training and simulation. By providing a more realistic and immersive virtual model, AR/VR technology can simulate different scenarios in a safe, controlled environment. This can reduce the risk of accidents and errors while improving user competence and performance (Lutters and Damgrave, 2019). Another important benefit of AR/VR-enabled DT is its ability to support remote collaboration. By allowing multiple users to visualise and interact with the virtual model in real time, AR/VR technology can strengthen remote collaboration between teams or organisations and improve communication, coordination, and shared understanding (Ait-Alla et al., 2019).

AR/VR technology is therefore critical in enabling DT because it provides a more immersive, intuitive, and engaging experience of the virtual model. XR-enabled DT can be used across different industries and offers important benefits such as training, simulation, design visualisation, and remote collaboration.

2.6.4 Cloud Computing

Cloud computing has been identified as a key enabler of DT because of its ability to store, process, and analyse large volumes of data in real time (Attaran and Celik, 2023). Cloud computing is a model for delivering computing services over the Internet, including data storage, processing power, and analytical capabilities. In the context of DT, cloud computing can be used to store and process the data generated by the physical asset, while also providing real-time analytics and actionable insights that support better decision-making (Hu et al., 2018). This makes cloud computing an important digital foundation for the effective deployment and operation of DT.

Cloud computing plays a central role in enabling DT by providing scalable, flexible infrastructure for data storage and processing. DT environments often involve large and continuous streams of data generated by sensors, control systems, and other connected devices. Cloud computing enables efficient storage and processing of this data while supporting real-time analytics and system responsiveness (Tao et al., 2019b). Cloud-computing-enabled DT can be applied across various industries, including manufacturing, energy, and healthcare (Tao et al., 2019e; Tao et al., 2018). In manufacturing, for example, cloud computing can be used to monitor and analyse the performance of machines and assembly lines and to optimise their operation in real time (Zhu, Liu and Xu, 2019). In energy and infrastructure systems, cloud computing can support the monitoring and control of plants and other facilities, thereby improving their performance, reliability, and operational efficiency (Attaran and Celik, 2023).

One of the key benefits of cloud-computing-enabled DT is its ability to support predictive maintenance. By processing and analysing data generated by the physical asset in real time, cloud computing can help the DT identify potential problems and predict their occurrence before they happen. This can reduce downtime, lower maintenance costs, and improve system reliability and performance (Tao et al., 2019a). Another important benefit of cloud-computing-enabled DT is its ability to support remote access and collaboration. By storing the virtual replica of the physical asset in the cloud, the DT can be accessed and interacted with by multiple users in real time from different locations. This improves communication and coordination among teams and organisations and also supports remote operation, supervision, and maintenance (Hu and Li, 2013).

Cloud computing is therefore critical to enabling DT because it provides the scalable, flexible, and accessible infrastructure needed for data storage, processing, and analytics. Cloud-computing-enabled DT can be used across different industries and offers important benefits such as predictive maintenance, remote access, and improved collaboration.

2.6.5 Big Data Analytics

Big data analytics has been recognised as a crucial enabler of DT due to its capacity to process and analyse large volumes of data in real time (Attaran and Celik, 2023). It involves examining extensive, complex, and diverse datasets to uncover hidden patterns, insights, relationships, and trends. In the context of DT, big data analytics can be utilised to process and analyse data generated by physical assets and to deliver real-time insights, forecasts, and decision support

(Tao et al., 2019b). This makes big data analytics a vital capability within the DT environment, especially when large, continuously evolving datasets need to be interpreted swiftly and accurately.

Big data analytics plays a crucial role in enabling DT by providing a platform for real-time data processing and analysis. Through big data analytics, anomalies, inefficiencies, and potential issues in the physical asset can be identified more easily, while useful insights and predictions can be generated more effectively (Tao and Zhang, 2017). This strengthens DT's ability to move beyond simple representation towards intelligent interpretation and proactive response. Big-data-analytics-enabled DT can be used in various industries, including manufacturing, healthcare, and transportation. In manufacturing, for example, big data analytics can be used to monitor and analyse the performance of machines and assembly lines and to predict possible issues before they occur. In healthcare, big data analytics can support more personalised treatment planning based on individual patient data and evolving health conditions (Attaran and Celik, 2023).

One of the major benefits of big data analytics for DT is its ability to support predictive maintenance. By analysing the data generated by the physical asset in real time, Big-Data-analytics-enabled DT can identify potential faults and predict their occurrence. This can reduce downtime, lower maintenance costs, and improve system performance and reliability (Wang et al., 2020). Another important benefit of Big-Data-analytics-enabled DT is its ability to support real-time decision-making. By generating real-time insights and predictions, big data analytics enables DT to deliver faster, more informed responses to changing operating conditions. This can help organisations respond more effectively to uncertainty, improve operational control, and strengthen performance across different systems (Zhu, Liu and Xu, 2019).

Big data analytics is therefore critical to enabling DT because it provides the analytical platform required for real-time data processing, interpretation, and insight generation. Big-Data-analytics-enabled DT can be applied in different industries and offers important benefits such as predictive maintenance and real-time decision-making.

2.7 Summary

This chapter provided a deep-dive into the concept of DT and its growing importance in the digital transformation of industries and systems. It began by explaining digital transformation as the broader shift from conventional operations to more connected, intelligent, and data-driven ways of working. In this context, the chapter distinguished between digitisation and digitalisation and showed how these developments have created the conditions for more advanced technologies such as the DT. The chapter then explored the concept of DT in detail, showing that it is more than a digital model. It is a dynamic virtual representation of a physical object, process, or system that can mirror real-world conditions, support analysis, and improve decision-making. Attention was also given to the composition of DT, including its physical entities, virtual models, data, services, and connections, as well as the basic requirements needed for successful deployment.

The discussion further examined the DT process structure through the Deming Cycle, illustrating how planning, implementation, evaluation, and continuous improvement support the system's effective functioning. The chapter also established CPS as a major stepping stone towards DT by explaining their concept, architecture, and contributions. Finally, the chapter examined the enabling technologies that make DT possible. These included the IoT, AI, extended reality, cloud computing, and big data analytics. Ultimately, the chapter showed that DT is a powerful and evolving technology that relies on integrating multiple digital systems and capabilities, and that it holds great potential for improving how complex systems are understood, managed, and optimised.

References

Ait-Alla, A. et al. (2019) 'Simulation-based analysis of the interaction of a physical and a Digital Twin in a cyber-physical production system', *IFAC MIM Proceedings*, 13(2008), pp. 1331–1336. doi:10.1016/j.ifacol.2019.11.383

Alam, K.M. and El Saddik, A. (2017) 'C2PS: A Digital Twin architecture reference model for the cloud-based cyber-physical systems', *IEEE Access*, 5, pp. 2050–2062. doi:10.1109/ACCESS.2017.2657006

Atlam, H.F. and Wills, G.B. (2020) 'IoT security, privacy, safety and ethics', in M. Farsi, A. Daneshkhah, A. Hosseinian-Far and H. Jahankhani (eds.), *Digital Twin technologies and smart cities*. Cham, Switzerland: Springer, pp. 123–149. doi:10.1007/978-3-030-18732-3_8

Attaran, M. and Celik, B.G. (2023) 'Digital Twin: Benefits, use cases, challenges, and opportunities', *Decision Analytics Journal*, 6, p. 100165. doi:10.1016/J.DAJOUR.2023.100165

Bhatti, G., Mohan, H. and Raja Singh, R. (2021) 'Towards the future of smart electric vehicles: Digital Twin technology', *Renewable and Sustainable Energy Reviews*, 141, p. 110801. doi:10.1016/J.RSER.2021.110801

Bloomberg, J. (2018) Digitization, digitalization, and digital transformation: Confuse them at your peril. *Forbes*. Available at: https://www.forbes.com/sites/jasonbloomberg/2018/04/29/digitization-digitalization-and-digital-transformation-confuse-them-at-your-peril/?sh=66b09ceb2f2c

Botkina, D. et al. (2018) 'Digital Twin of a cutting tool', in *51st CIRP Conference on Manufacturing Systems*, pp. 215–218. doi:10.1016/j.procir.2018.03.178

Cai, Y. et al. (2017) 'Sensor data and information fusion to construct digital-twins virtual machine tools for cyber-physical manufacturing', in *Proceedings of the 45th SME North American Manufacturing Research Conference (NAMRC)*, vol. 10, pp. 1031–1042. doi:10.1016/j.promfg.2017.07.094

Chapco-Wade-Safina, C. (2018) Digitization, digitalization, and digital transformation: What's the difference? *Medium*. Available at: https://colleenchapcowadesafina.medium.com/digitization-digitalization-and-digital-transformation-whats-the-difference-eff1d002fbdf

Dawkins, O., Hudson-Smith, A. and Dennett, A. (2018) 'Living with a Digital Twin: Operational management and engagement using IoT and mixed realities at UCL's here east campus on the Queen Elizabeth Olympic Park', *BIM-DigiCons*, p. 6. Available at: https://pdfs.semanticscholar.org/da61/a1bf345892d7504bde14624788ee9dd07b47.pdf (Accessed: 7 January 2020).

Fan, C. et al. (2019) 'Disaster City Digital Twin: A vision for integrating artificial and human intelligence for disaster management', *International Journal of Information Management*, p. 102049. doi:10.1016/j.ijinfomgt.2019.102049

Fortino, G. et al. (2016) 'Towards cyberphysical digital libraries: Integrating IoT smart objects into digital libraries', in A. Guerrieri, V. Loscri, A. Rovella and G. Fortino (eds.), *Management of cyber physical objects in the future Internet of Things*. Cham, Switzerland: Springer, pp. 135–156. doi:10.1007/978-3-319-26869-9_7

Gabor, T., Belzner, L., Kiermeier, M., Beck, M. T., and Neitz, A. (2016). 'A simulation-based architecture for smart cyber-physical systems', in *Proceedings of the IEEE International Conference on Autonomic Computing (ICAC)*. IEEE, pp. 374–379. doi:10.1109/ICAC.2016.29

Glaessgen, E.H. and Stargel, D.S. (2012) 'The Digital Twin paradigm for future NASA and U.S. Air force vehicles', in *Proceedings of the 53rd Structures, Structural Dynamics, and Materials Conference*. American Institute of Aeronautics and Astronautics, pp. 1–14. doi:10.2514/6.2012-1818

Grieves, M. and Vickers, J. (2016) 'Digital Twin: Mitigating unpredictable, undesirable emergent behavior in complex systems', in F.-J. Kahlen, S. Flumerfelt and A. Alves (eds.), *Transdisciplinary perspectives on complex systems: New findings and approaches*. Cham, Switzerland: Springer, pp. 85–113. doi:10.1007/978-3-319-38756-7_4

Hossain, M.A. and Nadeem, A. (2019) 'Towards digitizing the construction industry: State of the art of construction 4.0', in *Proceedings of the 10th International Structural Engineering and Construction Conference (ISEC)*. doi:10.14455/ISEC.RES.2019.184

Hribernik, K., Wuest, T. and Thoben, K.D. (2013) 'Towards product avatars representing middle-of-life information for improving design, development and manufacturing processes', *IFIP Advances in Information and Communication Technology*, 411, pp. 85–96. doi:10.1007/978-3-642-41329-2_10

Hu, L. et al. (2018) 'Modeling of cloud-based Digital Twins for smart manufacturing with MT connect', in *Proceedings of the 46th SME North American Manufacturing Research Conference (NAMRC)*. Elsevier, pp. 1193–1203. doi:10.1016/j.promfg.2018.07.155

Hu, Q. and Li, F. (2013) 'Hardware design of smart home energy management system with dynamic price response', *IEEE Transactions on Smart Grid*, 4(4), pp. 1878–1887. doi:10.1109/TSG.2013.2258181

Jakob, M. and Nilsson, P. (2018) *Virtual commissioning and construction of a Digital Twin for Smarta Fabriker*. Gothenburg: Chalmers University of Technology. Available at: https://odr.chalmers.se/items/b3cc4eac-ac48-4888-8b9bc872930ae3c2 (Accessed: 7 January 2020).

Kaewunruen, S. and Lian, Q. (2019) 'Digital Twin aided sustainability-based lifecycle management for railway turnout systems', *Journal of Cleaner Production*, 228, pp. 1537–1551. doi:10.1016/j.jclepro.2019.04.156

Kan, C. and Anumba, C.J. (2019) 'Digital Twins as the next phase of cyber-physical systems in construction', in *Proceedings of the ASCE International Conference on Computing in Civil Engineering*. Atlanta: American Society of Civil Engineers (ASCE), pp. 256–264. doi:10.1061/9780784482438.033

Kaur, M.J., Mishra, V.P. and Maheshwari, P. (2020) 'The convergence of Digital Twin, IoT, and machine learning: Transforming data into action', in M. Farsi, A. Daneshkhah, A. Hosseinian-Far and H. Jahankhani (eds.), *Digital Twin technologies and smart cities*. Cham, Switzerland: Springer, pp. 3–17. doi:10.1007/978-3-030-18732-3_1

Khajavi, S.H. et al. (2019) 'Digital Twin: Vision, benefits, boundaries, and creation for buildings', *IEEE Access*, 7. doi:10.1109/ACCESS.2019.2946515

Kraft, E.M. (2016) 'The US air force digital thread/Digital Twin – Life cycle integration and use of computational and experimental knowledge', in *Proceedings of the 54th AIAA Aerospace Sciences Meeting*. doi:10.2514/6.2016-0897

LaGrange, E. (2019, September) 'Developing a Digital Twin: The Roadmap for Oil and Gas Optimization,' in *Paper presented at the SPE Offshore Europe Conference and Exhibition*, Aberdeen, UK. doi:10.2118/195790-MS

Lee, E.A. (2008) 'Cyber physical systems: Design challenges', *Distributed Computing*, pp. 363–369. doi:10.1109/ISORC.2008.25

Lee, E.A. and Seshia, S.A. (2017) *Introduction to embedded systems - A cyber-physical systems approach*. Available at: https://ptolemy.berkeley.edu/books/leeseshia/releases/LeeSeshia_DigitalV2_2.pdf (Accessed: 8 July 2023).

Lee, J., Bagheri, B. and Kao, H.A. (2015) 'A cyber-physical systems architecture for Industry 4.0-based manufacturing systems', *Manufacturing Letters*, 3, pp. 18–23. doi:10.1016/J.MFGLET.2014.12.001

Legner, C. et al. (2017) 'Digitalization: Opportunity and challenge for the business and information systems engineering community', *Business and Information Systems Engineering*, 59(4), pp. 301–308. doi:10.1007/s12599-017-0484-2

Leitão, P.J. et al. (2015) 'Mapping beta diversity from space: Sparse generalised dissimilarity modelling (SGDM) for analysing high-dimensional data', *Methods in Ecology and Evolution*, 6(7), pp. 764–771. doi:10.1111/2041-210X.12378

Lontsikh, P.A. et al. (2020) 'Information security methods' application based on the digital management approaches and the Deming cycle in improving the modern production's processes', in *Proceedings of the IEEE International Conference "Quality Management, Transport and Information Security, Information Technologies" (IT and QM and IS)*. Yaroslavl, Russia: Institute of Electrical and Electronics Engineers Inc, pp. 123–126. doi:10.1109/ITQMIS51053.2020.9322856

Lu, Y. et al. (2020) 'Digital Twin-driven smart manufacturing: Connotation, reference model, applications and research issues', *Robotics and Computer-Integrated Manufacturing*, p. 101837. doi:10.1016/j.rcim.2019.101837

Lutters, E. and Damgrave, R. (2019) 'The development of Pilot Production Environments based on Digital Twins and virtual dashboards', in *Proceedings of the 29th CIRP Design*. Elsevier, pp. 94–99. doi:10.1016/j.procir.2019.04.228

Malakuti, S. et al. (2018) *Digital Twin – A key software component of Industry 4.0*. Ladenburg: ABB Review.

Mateev, M. (2020) 'Industry 4.0 and the Digital Twin for building industry', *International Scientific Journal "Industry 4.0"*, 5(1), pp. 29–32. Available at: https://stumejournals.com/journals/i4/2020/1/29.full.pdf (Accessed: 7 January 2020).

Min, Q. et al. (2019) 'Machine learning based Digital Twin framework for production optimization in petrochemical industry', *International Journal of Information Management*, 49, pp. 502–519. doi:10.1016/j.ijinfomgt.2019.05.020

Negri, E., Fumagalli, L. and Macchi, M. (2017) 'A review of the roles of Digital Twin in CPS-based production systems', *Procedia Manufacturing*, 11, pp. 939–948. doi:10.1016/J.PROMFG.2017.07.198

Ning, H.S. and Liu, H. (2015) 'Cyber-physical-social-thinking space based science and technology framework for the Internet of Things', *Science China Information Sciences*, 58(3), pp. 1–19. doi:10.1007/s11432-014-5209-2

Qi, Q. et al. (2018) 'Digital Twin service towards smart manufacturing', *Procedia CIRP*, 72, pp. 237–242. doi:10.1016/j.procir.2018.03.103

Qi, Q. et al. (2019) 'Enabling technologies and tools for Digital Twin', *Journal of Manufacturing Systems*, 58, pp. 3–21. doi:10.1016/j.jmsy.2019.10.001

Rabah, S. et al. (2018) 'Towards improving the future of manufacturing through Digital Twin and augmented reality technologies', in *Proceedings of the 28th International Conference on Flexible Automation and Intelligent Manufacturing (FAIM)*, pp. 460–467. doi:10.1016/j.promfg.2018.10.070

Rosen, R. et al. (2015) 'About the importance of autonomy and Digital Twins for the future of manufacturing', *IFAC-PapersOnLine*, pp. 567–572. doi:10.1016/j.ifacol.2015.06.141

Samir, K., Maffei, A. and Onori, M.A. (2019) 'Real-Time asset tracking; a starting point for Digital Twin implementation in manufacturing', in *Proceedings of the 52nd CIRP Conference on Manufacturing Systems*. Elsevier, pp. 719–723. doi:10.1016/j.procir.2019.03.182

Savić, D. (2019) 'From digitization, through digitalization, to digital transformation', in *Proceedings of the 21st International Conference on Grey Literature*, Hannover.

Scholz-Reiter, B. and Freitag, M. (2007) 'Autonomous processes in assembly systems', *CIRP Annals*, 56(2), pp. 712–729. doi:10.1016/J.CIRP.2007.10.002

Shafto, M. et al. (2010) 'DRAFT modeling, simulation, information technology & processing roadmap', *Technology Area*, 11, pp. 1–30.

Shahzad, M. et al. (2022) 'Digital Twins in built environments: An investigation of the characteristics, applications, and challenges', *Buildings*, 12(2), p. 120. doi:10.3390/BUILDINGS12020120

Singh, M. et al. (2022) 'Applications of Digital Twin across industries: A review', *Applied Sciences*, 12(11), p. 5727. doi:10.3390/APP12115727

Sleiti, A.K., Kapat, J.S. and Vesely, L. (2022) 'Digital Twin in energy industry: Proposed robust Digital Twin for power plant and other complex capital-intensive large engineering systems', *Energy Reports*, 8, pp. 3704–3726. doi:10.1016/J.EGYR.2022.02.305

Stark, R., Fresemann, C. and Lindow, K. (2019) 'Development and operation of Digital Twins for technical systems and services', *CIRP Annals*, 68(1), pp. 129–132. doi:10.1016/j.cirp.2019.04.024

Sun, X. et al. (2020) 'A Digital Twin-driven approach for the assembly-commissioning of high precision products', *Robotics and Computer-Integrated Manufacturing*, 61, pp. 1–14. doi:10.1016/j.rcim.2019.101839

Talkhestani, B.A. et al. (2018) 'Consistency check to synchronize the Digital Twin of manufacturing automation based on anchor points', in *Proceedings of the 51st CIRP Conference on Manufacturing Systems*. Elsevier, pp. 159–164. doi:10.1016/j.procir.2018.03.166

Tao, F. and Qi, Q. (2019) 'Make more Digital Twins', *Nature*. doi:10.1038/d41586-019-02849-1

Tao, F. and Zhang, M. (2017) 'Digital Twin shop-floor: A new shop-floor paradigm towards smart manufacturing', *IEEE Access*, 5, pp. 20418–20427. doi:10.1109/ACCESS.2017.2756069

Tao, F., Zhang, M. and Nee, A.Y.C. (2019c) 'Applications of Digital Twin', in B. Guerin (ed.), *Digital Twin driven smart manufacturing*. London: Academic Press, pp. 29–62. doi:10.1016/B978-0-12-817630-6.00002-3

Tao, F., Zhang, M. and Nee, A.Y.C. (2019d) 'Background and concept of Digital Twin', in B. Guerin (ed.), *Digital Twin driven smart manufacturing*. London: Academic Press, pp. 3–28. doi:10.1016/B978-0-12-817630-6.00001-1

Tao, F., Zhang, M. and Nee, A.Y.C. (2019e) 'Digital Twin, cyber–physical system, and Internet of Things', in B. Guerin (ed.), *Digital Twin driven smart manufacturing*. London: Academic Press, pp. 243–256. doi:10.1016/B978-0-12-817630-6.00012-6

Tao, F., Zhang, M. and Nee, A.Y.C. (2019f) 'Digital Twin and big data', in B. Guerin (ed.), *Digital Twin driven smart manufacturing*. London: Academic Press, pp. 183–202. doi:10.1016/B978-0-12-817630-6.00009-6

Tao, F., Zhang, M. and Nee, A.Y.C. (2019g) 'Digital Twin and cloud, fog, edge computing', in B. Guerin (ed.), *Digital Twin driven smart manufacturing*. London: Academic Press, pp. 171–181. doi:10.1016/B978-0-12-817630-6.00008-4

Tao, F., Zhang, M. and Nee, A.Y.C. (2019h) 'Digital Twin and services', in B. Guerin (ed.), *Digital Twin driven smart manufacturing*. London: Academic Press, pp. 203–217. doi:10.1016/B978-0-12-817630-6.00010-2

Tao, F., Zhang, M. and Nee, A.Y.C. (2019i) 'Digital Twin and virtual reality and augmented reality/mixed reality', in B. Guerin (ed.), *Digital Twin driven smart manufacturing*. London: Academic Press, pp. 219–241. doi:10.1016/B978-0-12-817630-6.00011-4

Tao, F. et al. (2018) 'Digital twin driven prognostics and health management for complex equipment', *CIRP Annals - Manufacturing Technology*, 67(1), pp. 169–172. doi:10.1016/j.cirp.2018.04.055

Tao, F. et al. (2019a) 'Digital Twins and cyber–physical systems toward smart manufacturing and Industry 4.0: Correlation and comparison', *Engineering*, 5(4), pp. 653–661. doi:10.1016/j.eng.2019.01.014

Tao, F. et al. (2019b) 'Digital Twin-driven product design framework', *International Journal of Production Research*, 57(12), pp. 3935–3953. doi:10.1080/00207543.2018.1443229

Thoben, K.D., Wiesner, S.A. and Wuest, T. (2017) '"Industrie 4.0" and smart manufacturing – A review of research issues and application examples', *International Journal of Automation Technology*, 11(1), pp. 4–16. doi:10.20965/IJAT.2017.P0004

Tuegel, E.J. et al. (2011) 'Reengineering aircraft structural life prediction using a Digital Twin', *International Journal of Aerospace Engineering*, 2011, pp. 1–14. doi:10.1155/2011/154798

Uhlemann, T.H.J., Lehmann, C. and Steinhilper, R. (2017a) 'The Digital Twin: Realizing the cyber-physical production system for Industry 4.0', in *Proceedings of the 24th CIRP Conference on Life Cycle Engineering*, pp. 335–340. doi:10.1016/j.procir.2016.11.152

Uhlemann, T.H.J. et al. (2017b) 'The Digital Twin: Demonstrating the potential of real time data acquisition in production systems', *Procedia Manufacturing*, 9, pp. 113–120. doi:10.1016/j.promfg.2017.04.043

Vathoopan, M. et al. (2018) 'Modular fault ascription and corrective maintenance using a Digital Twin', *IFAC-PapersOnLine*, 51(11), pp. 1041–1046. doi:10.1016/j.ifacol.2018.08.470

Vogel-Heuser, B. et al. (2017) 'Modularity and architecture of PLC-based software for automated production Systems: An analysis in industrial companies', *Journal of Systems and Software*, 131, pp. 35–62. doi:10.1016/J.JSS.2017.05.051

Wanasinghe, T.R. et al. (2020) 'Digital Twin for the oil and gas industry: Overview, research trends, opportunities, and challenges', *IEEE Access*, 8, pp. 104175–104197. doi:10.1109/ACCESS.2020.2998723

Wang, Y. et al. (2020) 'Big data driven hierarchical Digital Twin predictive remanufacturing paradigm: Architecture, control mechanism, application scenario and benefits', *Journal of Cleaner Production*, 248, p. 119299. doi:10.1016/j.jclepro.2019.119299

Woodhead, R., Stephenson, P. and Morrey, D. (2018) 'Digital construction: From point solutions to IoT ecosystem', *Automation in Construction*, 93, pp. 35–46. doi:10.1016/j.autcon.2018.05.004

Xu, Y. et al. (2018) 'Digital Twin concept in managing industrial capital construction projects life cycle', in *Proceedings of the 11th International Conference Management of Large-Scale System Development (MLSD)*. IEEE. doi:10.1109/MLSD.2018.8551867

Yu, W. et al. (2022) 'Energy Digital Twin technology for industrial energy management: Classification, challenges and future', *Renewable and Sustainable Energy Reviews*, 161, p. 112407. doi:10.1016/J.RSER.2022.112407

Zhou, C. et al. (2019) 'Cyber-physical-system-based safety monitoring for blind hoisting with the internet of things: A case study', *Automation in Construction*, 97, pp. 138–150. doi:10.1016/j.autcon.2018.10.017

Zhu, Z., Liu, C. and Xu, X. (2019) 'Visualisation of the Digital Twin data in manufacturing by using augmented reality', in *Proceedings of the 52nd CIRP Conference on Manufacturing Systems*, vol. 81, pp. 898–903. doi:10.1016/j.procir.2019.03.223

3 Application of Digital Twin in Various Industries

3.1 Introduction

Digital transformation is steadily changing how industries design, operate, monitor, and improve their systems. Within this shift, Digital Twin (DT) has emerged as one of the most influential technologies for linking the physical and digital worlds in a more intelligent and responsive manner. DT has attracted growing attention because it enables physical assets, processes, and systems to be represented in virtual form and continuously informed by real-world data. This makes it possible to simulate behaviour, analyse performance, predict faults, and support better decision-making across the lifecycle of assets and operations (Glaessgen and Stargel, 2012; Grieves and Vickers, 2016; Tao and Qi, 2019). Although the concept first gained widespread recognition in aerospace and manufacturing, it has now extended to many other economic industries.

The expanding use of DT is closely tied to the rise of Industry 4.0 and the growing integration of enabling technologies such as the Internet of Things (IoT), cloud computing, artificial intelligence (AI), big data analytics, and cyber-physical systems (CPS) (Lasi et al., 2014; Negri, Fumagalli and Macchi, 2017; Lu et al., 2020). These technologies have made it easier to gather real-time data, create intelligent virtual models, and maintain continuous communication between physical assets and digital systems. Consequently, DT has evolved beyond a visual model or simulation tool. It has become a strategic technology for enhancing efficiency, safety, reliability, maintenance, and innovation in increasingly complex operational environments (Kritzinger et al., 2018; Attaran and Celik, 2023).

The value of DT is especially evident in capital-intensive, technologically integrated industries that are highly dependent on performance, coordination, and lifecycle management. In such sectors, even minor inefficiencies or unexpected failures can result in high cost, operational disruption, or safety consequences. DT offers a way to reduce such risks by enabling organisations to observe system conditions more accurately, test possible interventions virtually, and optimise operations before acting in the physical environment (Tao, Zhang and Nee, 2019a; Singh et al., 2022). This explains why DT is

DOI: 10.1201/9781003743330-3

increasingly adopted across industries such as aerospace, automotive, manufacturing, oil and gas, mining, energy, and healthcare.

Another key reason for the rising interest in DT is its adaptability across various application areas. In some industries, DT is used to improve design and product development. In others, it supports predictive maintenance, production optimisation, safety planning, equipment monitoring, or personalised service delivery. This flexibility has established DT as a cross-sector technology with relevance extending far beyond its original industrial roots. Its proven usage across multiple economic sectors demonstrates that it can bolster both technical performance and strategic management, depending on the industry type and the complexity of the systems involved (Kan and Anumba, 2019; Shahzad et al., 2022; Attaran and Celik, 2023).

This chapter explores the application of DT in various economic industries. It begins with the aerospace industry and then discusses applications in automotive, manufacturing, oil and gas, mining, energy, and healthcare. It also considers other economic sectors where DT is gaining relevance and concludes by discussing the documented use of DT within the existing body of knowledge. In doing so, the chapter shows that DT is not confined to a single field but is increasingly a transformative technology across diverse industries and operational settings.

3.2 Digital Twin in the Aerospace Industry

DT has gained considerable traction across several industries, and the aerospace industry remains one of its most prominent applications. This is largely because aerospace systems are highly complex, safety-critical, and expensive to design, operate, and maintain. In such an environment, there is a strong need for technologies that can improve accuracy, reduce uncertainty, support lifecycle management, and enhance decision-making. DT addresses these needs by creating a virtual representation of a physical aircraft, spacecraft, subsystem, or operational process, which can then be used to mirror, analyse, and optimise the behaviour of the physical counterpart throughout its lifecycle (Glaessgen and Stargel, 2012; Tuegel et al., 2011; Tao, Zhang and Nee, 2019a).

The aerospace industry was among the earliest sectors to embrace DT because of its long-standing dependence on modelling, simulation, and system-based engineering. In aerospace applications, DT is not limited to representing the external form of an aircraft or spacecraft. It also captures the condition, behaviour, and interaction of multiple systems and subsystems, including propulsion, structural performance, electrical systems, thermal control, and operational environments. This makes it possible to assess system behaviour more accurately before physical deployment and to continue monitoring performance after deployment. In this sense, DT supports a continuous relationship between design, manufacturing, operation, maintenance, and improvement, thereby making it highly valuable in an industry where failures can have serious financial and safety implications (Kraft, 2016; Li et al., 2022; Xiong and Wang, 2022).

A notable example is the use of DT by the National Aeronautics and Space Administration (NASA) in the development and improvement of its next-generation spacecraft, the Orion Multi-Purpose Crew Vehicle. NASA uses DT to simulate key systems, such as propulsion, electrical power, and thermal control, so that potential issues can be identified, analysed, and resolved before physical implementation (Li et al., 2022). This illustrates one of the major strengths of DT in aerospace: the ability to test, predict, and refine system performance in a virtual environment before committing to costly real-world execution.

More broadly, DT is helping the aerospace industry move from conventional, largely reactive approaches to more intelligent, predictive, and lifecycle-oriented modes of operation. It provides a platform through which design decisions can be tested, operating conditions can be simulated, maintenance can be predicted, and performance can be continuously improved. As aerospace systems become more advanced and operational demands become more stringent, DT is increasingly being recognised as a strategic technology for improving reliability, efficiency, safety, and innovation across the sector (Phanden, Sharma and Dubey, 2021; Li et al., 2022; Xiong and Wang, 2022).

3.2.1 Benefits of DT in the Aerospace Industry

The benefits of DT in the aerospace industry are substantial. One of the most important is its ability to simulate different scenarios and optimise aircraft performance before physical implementation. This can support reductions in fuel consumption, improvements in operational safety, and enhancements in passenger comfort and system reliability (Li et al., 2022). Because the aerospace industry operates in highly demanding environments, the ability to test and refine systems virtually before deployment provides a major advantage.

DT also improves the efficiency of aircraft maintenance and repair. By creating a virtual model of an aircraft and linking it to real-time operational data, engineers can continuously monitor performance and identify early signs of wear, malfunction, or system failure. This allows maintenance needs to be detected before they become major problems. As a result, downtime can be reduced, maintenance costs can be lowered, and aircraft safety can be improved (Xiong and Wang, 2022). In this sense, DT supports a more proactive and predictive maintenance culture within the aerospace industry.

Another major benefit of DT is its ability to accelerate the design and manufacturing process. Using virtual models, engineers can quickly and efficiently test different designs, configurations, and performance conditions without relying solely on physical prototypes. This reduces the time and cost associated with development and testing while improving confidence in the final design (Phanden, Sharma and Dubey, 2021). In an industry where development cycles are long and highly expensive, this advantage is especially valuable.

3.2.2 Challenges of DT in the Aerospace Industry

Despite its many benefits, the use of DT in the aerospace industry is not without challenges. One of the major challenges is data integration. Aerospace systems generate large volumes of data from different sources, including sensors, onboard systems, engineering models, and operational records. Integrating these different data types into a coherent, functional DT environment can be highly complex and time-consuming. The quality, consistency, and compatibility of data remain important concerns in this regard (Li et al., 2022).

Another significant challenge relates to cybersecurity. DT depends on the continuous transfer, storage, and analysis of sensitive operational and engineering data. In the aerospace industry, such data are often critical and highly confidential. This means that strong cybersecurity measures are required to protect the system's integrity, privacy, and reliability. Weaknesses in data protection can expose the system to serious operational and security risks (Li et al., 2022; Xiong and Wang, 2022).

3.3 Digital Twin in the Automotive Industry

DT is increasingly being adopted across different industries for its potential to optimise operations, improve productivity, and reduce costs. One of the industries where its relevance is becoming more pronounced is the automotive industry. This is largely because modern vehicles are highly complex systems composed of tightly integrated mechanical, electrical, electronic, and software-based components. As vehicles become more intelligent, connected, and automated, the need for technologies to support better design, monitoring, analysis, and optimisation has also increased. In this context, DT provides an important digital framework through which manufacturers can better understand and manage vehicle systems and production processes (Tao, Zhang and Nee, 2019a; Attaran and Celik, 2023).

In the automotive industry, a DT of a vehicle refers to a virtual representation of a physical vehicle that integrates data from multiple sensors, components, and systems. This virtual model can mirror the vehicle's condition, behaviour, and performance in real time or near real time. It can also simulate various scenarios, environments, and operating conditions, enabling engineers to identify and address potential issues before they occur in the physical world (Attaran and Celik, 2023). This capability is highly valuable in the automotive sector, where design decisions, system interactions, and performance outcomes must be carefully assessed before mass production and market deployment.

The use of DT in the automotive industry extends across the entire vehicle lifecycle, from design and development to manufacturing, operation, maintenance, and user experience. In product design and development, DT allows engineers to create a virtual version of the vehicle early in the design process. This makes it possible to simulate the performance of different systems and components, test alternative designs, and identify weaknesses before physical prototypes are built. Such an approach can reduce development time, lower design costs, and improve the vehicle's safety, reliability, and performance (Damjanovic-Behrendt, 2018). It also supports a more iterative and informed design process, where modifications can be made virtually before they are introduced into the physical product.

DT is also being used in automotive manufacturing and production processes. By creating a DT of a production line, assembly system, or factory environment, engineers and production managers can simulate different production scenarios and optimise workflow, resource use, and machine coordination. This can improve efficiency, reduce waste, minimise downtime, and strengthen overall competitiveness in the market (Piromalis and Kantaros, 2022). In this sense, DT not only supports the vehicle itself but also supports the industrial systems through which the vehicle is produced.

In addition, DT is increasingly relevant to vehicle operation and maintenance. Automakers can use DT to simulate vehicle behaviour in virtual environments before physical deployment. In contrast, operational data from vehicles in use can help improve performance assessment, fault detection, and maintenance planning. This supports a more predictive and proactive approach to vehicle management. As a result, DT can help improve product quality, reduce maintenance costs, and increase customer satisfaction over the life of the vehicle (Bhatti, Mohan and Raja Singh, 2021; Attaran and Celik, 2023). The automotive industry is therefore increasingly recognising DT as a strategic tool for improving not only product development but also operational efficiency and long-term value creation.

3.3.1 Benefits of DT in the Automotive Industry

DT offers several important benefits to the automotive industry. One of the most significant is the reduction in development time. By enabling the simulation of vehicle performance and system behaviour before physical prototypes are produced, DT allows manufacturers to identify problems earlier and refine designs more efficiently. This supports faster innovation and more cost-effective development processes (Damjanovic-Behrendt, 2018). DT also improves product performance and quality by enabling engineers to evaluate the interactions among different components and systems in virtual environments before physical deployment. This helps to improve safety, reliability, and functionality.

Another major benefit is improved manufacturing efficiency. By modelling production lines and factory processes, DT enables manufacturers to optimise operations, reduce waste, and minimise downtime (Piromalis and Kantaros, 2022). DT also supports predictive maintenance by helping manufacturers and operators detect issues early and plan interventions more effectively. This can reduce maintenance costs and improve the lifespan of assets and vehicles. In addition, DT can improve customer satisfaction by supporting better product reliability, more efficient servicing, and enhanced vehicle performance over time (Bhatti, Mohan and Raja Singh, 2021).

3.3.2 Challenges of DT in the Automotive Industry

Despite its many advantages, implementing DT in the automotive industry also presents several challenges. One of the major challenges is data management. DT generates a large volume of data that must be stored, processed, integrated, and analysed continuously. Managing this data effectively can be complex and often requires substantial investment in software and hardware infrastructure (Piromalis and Kantaros, 2022). Another challenge is data security. The automotive industry handles sensitive information relating to product design, production systems, and customer data. Protecting this information is critical, and manufacturers must implement strong cybersecurity measures to reduce the risk of cyber-attacks and data breaches (Piromalis and Kantaros, 2022).

Integration with legacy systems is also a challenge. Many automakers still rely on legacy systems that may not readily integrate with DT technologies. Integrating DT with such systems can be difficult and may require major upgrades in infrastructure and processes (Bhatti, Mohan and Raja Singh, 2021). Finally, cost remains an important concern. The implementation of DT requires significant investment in digital infrastructure, software, hardware, and specialised expertise. For smaller automakers, this can present a major barrier. In addition, the maintenance and upgrading of DT-related infrastructure can require ongoing investment over time (Bhatti, Mohan and Raja Singh, 2021).

3.4 Digital Twin in the Manufacturing Industry

DT is transforming a wide range of industries, and the manufacturing industry remains one of its most important areas of application. This is largely because manufacturing environments are highly process-driven, data-intensive, and dependent on precision, coordination, and continuous performance improvement. In recent years, manufacturers have adopted DT to improve manufacturing processes, increase efficiency, reduce cost, and enhance product quality (Lu et al., 2020). In manufacturing, a DT is generally understood as a virtual representation of a physical object, machine, production line, or complete factory system, used to simulate its behaviour, performance, condition, and maintenance requirements. A DT can therefore be created for a product, an item of equipment, a production cell, a full assembly line, or even an entire plant. It is an accurate digital replica of a physical asset or process, developed through data gathered from sensors, equipment, and other operational sources (Tao et al., 2019a).

The manufacturing industry has been one of the strongest drivers of DT development because it already depends heavily on automation, production monitoring, system integration, and continuous optimisation. In such an environment, DT provides a valuable means of linking the physical shop floor with a digital environment where processes can be observed, analysed, tested, and improved. This allows manufacturers to move beyond conventional monitoring and reactive control towards more intelligent and predictive forms of production management. Through DT, manufacturers can better understand how machines, processes, and products behave over time and under changing operating conditions. This improves their ability to manage complexity, reduce inefficiencies, and respond quickly to disruptions.

A major strength of DT in manufacturing lies in its ability to mirror production processes in real time or near real time. By integrating physical data with a virtual model, manufacturers can simulate production performance, evaluate alternative scenarios, and make better operational decisions before implementing changes in the physical environment. This creates opportunities for process optimisation, better scheduling, more efficient use of resources, and improved coordination across the production system (Kritzinger et al., 2018; Attaran and Celik, 2023). It also supports a more comprehensive understanding of the relationships among machine performance, workflow, product quality, and maintenance requirements. As a result, DT becomes more than a digital replica. It becomes a management and decision-support tool that improves the design, operation, and refinement of manufacturing systems.

The use of DT in manufacturing also extends across the full production lifecycle. During product development, DT can be used to model and test how a product is expected to perform before physical production begins. During manufacturing operations, it can monitor machine conditions, production flow, and system interactions. During maintenance, it can support fault detection, performance analysis, and intervention planning. This broad applicability makes DT especially valuable in manufacturing, where decisions in one stage of the process often influence performance in another. By connecting data, models, and decision-making within one digital environment, DT supports a more integrated, responsive, and efficient manufacturing system.

3.4.1 Benefits of DT in the Manufacturing Industry

The use of DT in manufacturing offers several important benefits, including improved production efficiency, reduced downtime, lower cost, and enhanced product quality. One important benefit is predictive maintenance. DT can predict when a machine is likely to fail or require maintenance, enabling manufacturers to schedule maintenance before a breakdown occurs. This can reduce unplanned downtime and improve machine availability (Rabah et al., 2018). Another important benefit is enhanced product quality. DT can be used to simulate a product's performance and behaviour before it is manufactured. This allows manufacturers to identify possible defects, weaknesses, or quality issues early in the process. Such an approach can reduce the number of defective products and improve customer satisfaction (Zhang et al., 2019).

DT also supports improved efficiency. It can be used to optimise production schedules, reduce bottlenecks, and improve production flow. This helps manufacturers to produce goods faster, at lower cost, and with fewer resources (Tao, et al., 2019). In addition, DT helps reduce costs by lowering expenses associated with maintenance, repairs, and downtime. By predicting failures before they occur, manufacturers can schedule interventions at more convenient times and avoid the high cost of unexpected production interruptions (Kunath and Winkler, 2018).

3.4.2 Challenges of DT in the Manufacturing Industry

Despite its many advantages, the adoption of DT in manufacturing also presents several challenges. One major challenge is data integration. Creating a DT requires data from many different sources, including sensors, equipment, and other production systems. Integrating this data can be difficult because it often comes from different platforms and in different formats (Kunath and Winkler, 2018). Another challenge is data security. DT relies on large volumes of operational data, which may be vulnerable to cyber-attacks or unauthorised access. Ensuring the security of DT data is critical because any breach can disrupt production and compromise system reliability (Attaran and Celik, 2023).

A further challenge is the high cost of development and deployment. Building a DT often requires major investment in hardware, software, data infrastructure, and technical expertise. This can be difficult for smaller manufacturers and may limit wider adoption (Kunath and Winkler, 2018). There is also the challenge of integration with existing systems. Many manufacturing organisations still rely on older systems and established workflows that may not readily align with DT technology. Integrating DT into such environments may require major changes in systems, processes, and staff practices, which can sometimes result in resistance from employees accustomed to traditional manufacturing methods (Qi et al., 2018).

3.5 Digital Twin in the Oil, Gas, Mining, and Energy Industries

DT has emerged as a powerful tool in the oil, gas, mining, and energy industries for improving operational efficiency, reliability, safety, and decision-making. These industries are characterised by large-scale assets, hazardous environments, high operational costs, and complex system interactions. For this reason, there is a strong need for technologies that can support continuous monitoring, predictive analysis, and process optimisation. DT responds to this need by creating a virtual model of a physical asset, system, or process designed to mirror the behaviour and condition of its real-world counterpart. This allows operators to observe performance, identify inefficiencies, predict failures, and test alternative strategies before implementing them in the physical environment (Tao et al., 2019a; Attaran and Celik, 2023).

The value of DT in these industries lies in its ability to integrate real-time data, historical records, and simulation capabilities into one digital environment. In oil and gas operations, this can include the monitoring of drilling systems, pipelines, refineries, and offshore platforms. In mining, DT can be applied to equipment, underground systems, haulage operations, ventilation systems, and mineral processing activities. In the energy sector, DT can be used for power plants, transmission systems, renewable energy systems, and other infrastructure. Through this integration, DT provides operators with a clearer and more dynamic understanding of how systems perform under changing operational conditions. It also supports more proactive management by enabling decisions based on real-time system intelligence rather than on delayed or fragmented information.

One of the most important applications of DT in the mining and energy industries is the monitoring and maintenance of equipment. By creating a DT of a piece of equipment, operators can monitor its condition and performance in real time, predict when maintenance will be required, and identify potential faults before they become serious problems. This helps to reduce downtime, extend the useful life of equipment, and improve operational safety by detecting hazards early (Yu et al., 2022). In sectors where equipment failure can result in major production losses, environmental damage, or safety incidents, this predictive capability is especially valuable. DT is also being used to optimise the performance of energy systems. By creating a virtual model of a power plant or energy network, operators can simulate different operating conditions and determine the most efficient ways to generate, distribute, and manage energy. This can help reduce costs, improve supply reliability, and identify opportunities for cleaner, more efficient energy use. It also strengthens operators' ability to evaluate different scenarios without disrupting the physical system (Sleiti, Kapat and Vesely, 2022). In this way, DT supports both operational efficiency and long-term sustainability objectives.

Beyond performance optimisation, DT also contributes significantly to safety management. The oil, gas, mining, and energy industries often operate in environments where accidents can have severe human, environmental, and financial consequences. By using DT to simulate dangerous scenarios, operators can identify safety risks, test emergency responses, and develop more effective mitigation strategies before such events occur in the real world. For example, a DT for a refinery may be used to simulate emergencies such as fire, explosion, or system failure to determine the most appropriate response (Wanasinghe et al., 2020). This ability to rehearse and analyse hazardous scenarios in a virtual environment makes DT especially valuable in high-risk industries.

The application of DT in these sectors is also expanding beyond asset-level monitoring towards broader system optimisation. It is increasingly being used to improve resource management, reduce waste, optimise supply chains, and support more sustainable operations. As digitalisation continues to advance, DT is becoming an important strategic tool for improving both day-to-day operations and long-term planning in oil, gas, mining, and energy systems.

3.5.1 Benefits of DT in the Oil, Gas, Mining, and Energy Industries

The benefits of DT in the mining, oil, gas, and energy industries are considerable. One important benefit is real-time monitoring and control. DT enables the real-time monitoring of equipment and processes, allowing problems to be identified and resolved quickly. This reduces downtime and improves productivity (Singh et al., 2022). Another major benefit is improved safety. DT can simulate dangerous scenarios, such as underground collapses, gas leaks, fires, or explosions, thereby allowing operators to test emergency response plans without exposing workers to danger (El Bazi et al., 2023).

DT also supports predictive maintenance. By continuously monitoring equipment performance, the DT can predict when maintenance is likely to be needed. This reduces unplanned downtime and lowers maintenance costs (Yu et al., 2022). Another important benefit is reduced cost. DT can optimise processes, reduce waste, and improve resource utilisation, thereby leading to significant cost savings (Sleiti, Kapat and Vesely, 2022). In addition to these benefits, DT also has future potential in autonomous operations, sustainability improvement, resource optimisation, and supply chain optimisation. It can be used to simulate and improve autonomous mining operations, optimise energy use and waste management, reduce resource loss, and improve the efficiency of the full supply chain from extraction to transport and processing (Chen et al., 2021; El Bazi et al., 2023; Yu et al., 2022; Wanasinghe et al., 2020).

3.5.2 Challenges of DT in the Oil, Gas, Mining, and Energy Industries

Despite its many benefits, DT in the oil, gas, mining, and energy industries also presents several challenges. One challenge is DT complexity. Creating and maintaining a DT requires a complex technical architecture and skilled personnel with expertise in data analysis, modelling, and simulation. This complexity can make implementation difficult and resource-intensive (Sleiti, Kapat and Vesely, 2022; Attaran and Celik, 2023).

Another challenge is data management. DT relies on large volumes of real-time data collected from multiple sources, including sensors, equipment, and operational processes. Collecting, processing, and analysing this data effectively remains a major challenge. Security is also a serious concern. Because DT relies on sensitive operational and technical data, secure storage and management are essential to prevent unauthorised access, cyber-attacks, and data breaches. In addition, DT requires a strong infrastructure. A reliable and robust IT environment is needed to support the large volume of real-time data involved. There is also the challenge of integration, as DT requires coordinating multiple technologies, including IoT, AI, and cloud computing, to create a seamless, functional system capable of real-time monitoring and optimisation.

3.6 Digital Twin in the Healthcare Industries

DT has continued to gain momentum across different industries, and the healthcare industry is no exception. Healthcare remains one of the most critical sectors of society because it is directly linked to human life, well-being, and quality of care. For this reason, the introduction of intelligent digital technologies into healthcare has attracted growing attention. Among these technologies, DT has emerged as a promising tool for improving patient outcomes, strengthening healthcare management, and supporting more informed clinical decision-making (Attaran and Celik, 2023). In the healthcare context, a DT refers to a virtual model that simulates the human body, a bodily system, an organ, a patient-specific condition, or even a medical device. This virtual model is created by combining patient data from multiple sources, including electronic health records (EHRs), medical imaging, sensor-based monitoring systems, and wearable devices. Once developed, the model can be used to run simulations, predict outcomes, and support the design of personalised treatment strategies. The goal of DT in healthcare is to provide clinicians with a more comprehensive understanding of a patient's condition and to support early diagnosis, targeted interventions, and improved treatment outcomes (Erol, Mendi and Dogan, 2020).

The relevance of DT in healthcare lies in its ability to move beyond generalised treatment approaches towards more patient-specific and data-driven care. Traditional healthcare systems often depend on periodic assessments, standard treatment protocols, and reactive intervention. By contrast, DT offers the possibility of continuous insight into patient conditions and system performance. Through simulation and predictive analysis, clinicians can understand how a disease may progress, how a patient may respond to treatment, or how a medical device may perform under different conditions. This creates opportunities for more precise, proactive, and efficient healthcare delivery. DT is also important because healthcare systems are increasingly complex. They involve patients, clinicians, devices, hospital infrastructure, treatment protocols, data platforms, and regulatory systems, all of which interact in ways that affect outcomes. By creating virtual representations of patients or devices, DT allows these interactions to be studied and managed more effectively. In this way, DT is not only useful for clinical treatment but also for equipment management, training, and healthcare operations more broadly. Its growing role in healthcare reflects the sector's wider digital transformation and the need for smarter tools to support safer, more personalised, and more responsive care.

3.6.1 Benefits of DT in the Healthcare Industries

The benefits of DT in healthcare are wide-ranging. One of the most important is personalised treatment. DT can help healthcare professionals develop patient-specific treatment plans by simulating different scenarios and predicting the most suitable course of action. This can lead to more effective and efficient treatment while reducing the cost of care (Augustine, 2020). Another important benefit is improved diagnostics. DT can support the simulation of different disease conditions and likely outcomes, which can help clinicians reach earlier and more accurate diagnoses. This increases the likelihood of timely intervention and improves clinical outcomes (Bende and Gochhait, 2020). DT also supports predictive maintenance of medical devices. By simulating device performance and monitoring their condition, DT can indicate when equipment is likely to fail or require servicing. This can reduce downtime and minimise the risk of equipment-related disruption in patient care (Augustine, 2020).

DT also enables remote monitoring. Through data from connected devices and digital systems, patients can be monitored outside the hospital environment while healthcare providers receive real-time feedback. This can reduce hospital readmissions, lower the cost of care, and improve patient management and outcomes (Jimenez, Jahankhani and Kendzierskyj, 2020). In addition, DT supports medical education. It can be used to simulate disease conditions and medical procedures, thereby helping healthcare professionals to improve their knowledge, skills, and understanding of the human body and clinical practice (Hassani, Huang and MacFeely, 2022).

3.6.2 Challenges of DT in the Healthcare Industries

Despite its promise, the use of DT in healthcare also presents important challenges. One major challenge is data privacy. DT relies on the collection, storage, and use of large volumes of patient data, raising serious concerns about confidentiality, ethical use, and cybersecurity. Healthcare providers must ensure that patient data are protected and handled responsibly (Jimenez, Jahankhani and Kendzierskyj, 2020; Hassani, Huang and MacFeely, 2022). Another challenge is regulatory compliance. The use of DT in healthcare is subject to regulations, standards, and legal requirements, which may vary across countries and jurisdictions. Healthcare organisations must therefore ensure that DT implementation complies with the relevant legal and ethical frameworks (Hassani, Huang and MacFeely, 2022). A further challenge is technical complexity. The development, implementation, and maintenance of DTS in healthcare can be highly complex and may require specialised expertise, infrastructure, and interdisciplinary collaboration. Healthcare providers must therefore ensure that the technical capacity needed for successful deployment is available (Ahmed and Devoto, 2020).

3.7 Documented Practical Use of Digital Twin within the Existing Body of Knowledge

Since the beginning of the Third Industrial Revolution, technological innovation has continued to transform industrial productivity and reshape how systems are designed, operated, and managed. In recent decades, this transformation has become more apparent through digitalisation and the rise of intelligent technologies, emphasising what is now widely known as digital transformation. One of the most recognised expressions of this shift is Industry 4.0, a term introduced in 2011 to promote more computerised, connected, and intelligent production systems. The Fourth Industrial Revolution is now visible across various sectors, including manufacturing, energy, logistics, and transportation (Lasi et al., 2014). Within this broader transformation, DT has emerged as a key technological concept, supported by related innovations such as cloud computing, big data, the IoT, and CPS (Negri, Fumagalli and Macchi, 2017).

Although DT is often presented as a modern digital innovation, the core idea predates the term itself. The concept of a "twin" can be linked to earlier efforts to replicate physical systems for simulation and operational support. Rosen et al. (2015) traced this idea back to the Apollo missions, where mirrored spacecraft systems were used for training and mission assistance. From a more formal scholarly standpoint, Grieves and Vickers (2016) established the conceptual foundations of DT within product lifecycle management, demonstrating how physical products and their digital counterparts could be more closely connected across the entire lifecycle. In this sense, DT did not emerge in isolation. Rather, it evolved from earlier attempts to enhance visibility, simulation, monitoring, and lifecycle decision-making within complex systems. Its early application in aerospace was followed by broader use in manufacturing, robotics, industrial engineering, and informatics (Negri, Fumagalli and Macchi, 2017).

What is significant in the current body of knowledge is that DT has moved beyond purely conceptual discussions and has increasingly been demonstrated in practical and applied settings. Table 3.1 illustrates this progression by showing how DT has been adopted across various academic and industrial contexts. The table clearly indicates that DT is no longer limited to a single discipline or type of system. Rather, it has been embraced in an expanding range of sectors where complex physical assets, real-time data, and lifecycle decisions must be managed together. In this regard, the practical value of DT lies in its ability to enhance system visibility, enable predictive maintenance, support decision-making, optimise operations, and provide lifecycle intelligence.

Table 3.1 Academic Documentation of the Application of DT

Academic Study Reference	*Field of Study*	*DT Application*
Tuegel et al. (2011)	Aerospace	Aircraft structural life prediction using a tail-number-specific computational twin to forecast damage progression and remaining useful life.
Glaessgen and Stargel (2012)	Aerospace	NASA/United States Air Force (USAF) DT paradigm linking high-fidelity models, IVHM sensor updates, and fleet maintenance data to assess system health and project remaining useful life.
Rosen et al. (2015)	Manufacturing	DT concepts are used to support more autonomous production systems, faster response to disturbances, and improved operational decision-making.

(*Continued*)

Table 3.1 Continued

Academic Study Reference	*Field of Study*	*DT Application*
Grieves and Vickers (2016)	Product lifecycle management	DT is applied to reduce unsafe and unpredictable emergent behaviour in complex systems through improved visibility, monitoring, and lifecycle decision-making.
Canedo (2016)	Informatics/Industrial Internet of Things (IIoT)	DT and IIoT lifecycle optimisation through industrial information flows that support asset and process lifecycle management.
Kraft (2016)	Aerospace	USAF Digital Thread/DT lifecycle integration combining computational and experimental knowledge to support fleet decisions across lifecycle phases.
Schluse and Rossmann (2016)	Robotics/systems engineering	Experimentable DTs used for virtual commissioning and simulation-based development of complex technical systems in engineering and robotics contexts.
Alam and El Saddik (2017)	Informatics	DT architecture for CPS integration across domains through different interaction modes.
Li et al. (2017)	Aerospace	Aircraft wing health monitoring and crack growth prediction through a DT prototype that informs decision-makers on damage extent.
Schleich et al. (2017)	Manufacturing	Reference modelling approach for managing geometric variation through a scalable and interoperable DT in production engineering.
Uhlemann, et al. (2017)	Manufacturing	DT-enabled cyber-physical manufacturing system for greater transparency and near-real-time monitoring and control.
Söderberg et al. (2017)	Industrial engineering	Individualised production supported by real-time geometry verification through DT-based quality control.
Tao, Qi, et al. (2019)	Product lifecycle management	DT is applied across design, manufacturing, and service using big data to improve lifecycle intelligence and sustainability.
Leng et al. (2019)	Manufacturing	DT and CPS are used to monitor and control a smart workshop environment.
Borth, Verriet, and Muller (2019)	Built environment	DT and CPS concepts applied to support architectural design and development workflows.

(*Continued*)

Table 3.1 Continued

Academic Study Reference	*Field of Study*	*DT Application*
Aivaliotis et al. (2019)	Manufacturing/ maintenance	DT-enabled predictive maintenance and remaining useful life workflow demonstrated in an industrial maintenance context.
Feldt et al. (2020)	Manufacturing	DT is used in a manufacturing company to enable and assess the potential for real-time autonomous decision-making.
Platenius-Mohr et al. (2020)	Manufacturing	IIoT DT interoperability achieved through model transformation using an Asset Administration Shell case.
Marmolejo-Saucedo (2020)	Pharmaceutical manufacturing	Design and development of a DT within a pharmaceutical manufacturing company case study.
Fotland, Haskins, and Rølvåg (2020)	Lifecycle management	DT applied to work-environment safety and crane cable-pulley systems in an industrial safety context.
Ngwenya, Issa, and Anumba (2021)	Finance/built environment	The DT approach is used to visualise building costs across the lifecycle and improve lifecycle cost visibility.
Hassani, et al. (2022)	Healthcare	Sector-wide exploration of DT use in healthcare.
Madubuike and Anumba (2022)	Healthcare/facility management	DT-aided healthcare facility management case study at Shanghai Tongji Hospital, linking facility information with operational decision-making.
Kandasamy et al. (2022)	Energy/smart-grid cybersecurity	Electric power DT is developed for a physical smart-grid testbed to support cybersecurity testing experiments.
Steyn and Broekman (2022)	Transport infrastructure	DT of a local road network developed for monitoring and managing road infrastructure behaviour.
Mendi (2022)	Production line	Automotive production line DT using Message Queuing Telemetry Transport to monitor performance and reduce manufacturing downtime.
Doubell (2023)	Rail/transport infrastructure	DT applied to railway infrastructure in the Passenger Rail Agency of South Africa context to support maintenance and asset management.
Ashrafian and Pedersen (2023)	Logistics/warehousing	DT is developed for a complex logistics and automated-warehousing case, focusing on order-picking and operational performance.

(*Continued*)

Table 3.1 Continued

Academic Study Reference	*Field of Study*	*DT Application*
Mrzyk et al. (2023)	Energy/industrial maintenance	Predictive maintenance DT architecture applied to a compressor and adapted to a wind turbine use case.
Wang et al. (2024)	Oil and gas/pipeline integrity	DT-oriented pipeline condition monitoring for integrity assessment and decision support in field or prototype contexts.
Chacón et al. (2024)	Construction/built environment	DT of an office building construction process using site-data pipelines and dashboards for monitoring and decision support.
Arsecularatne et al. (2024)	Building energy/ facilities	DT-driven approach to reduce building energy use by modelling occupant behaviour and enabling operational interventions.
Pacheco-Blazquez et al. (2024)	Offshore renewable energy	DT applied to assess the remaining useful life of offshore wind turbine structures for reliability-related decisions.
Ma et al. (2025)	Building operations	Mixed reality-assisted DT implemented in a real office building, integrating Building Information Modelling (BIM), Building Automation Systems (BAS)/IoT, AI, and mixed reality for Heating, Cooling, and Air Conditioning (HVAC) and indoor environmental monitoring and control.
Hwang et al. (2025)	Building energy management	DT-driven building energy management system applied in an operational BEMS context.
Machalski et al. (2025)	Energy (hydropower)	DT is implemented for a hydroelectric power plant to support monitoring, maintenance, and operations.
Nasim et al. (2025)	Civil infrastructure (bridges)	DT framework is demonstrated for bridge monitoring by integrating sensing and inspection data to support lifecycle decision-making.
Park et al. (2025)	Water infrastructure	DT tool developed for drinking water distribution networks with augmented reality (AR)-enabled interaction and decision support.

The earliest documented practical applications are strongly rooted in aerospace. Tuegel et al. (2011) applied a computational twin to aircraft structural life prediction, while Glaessgen and Stargel (2012) advanced the NASA and USAF DT paradigm by linking high-fidelity models with sensor updates and maintenance-related information. Kraft (2016) also demonstrated how the USAF used the Digital Thread and DT concepts to support lifecycle integration and fleet-level decisions. These early aerospace applications are important because they showed the practical value of DT in safety-critical, capital-intensive, and highly complex systems. They also helped establish one of the most enduring themes in DT applications: the use of digital representation to improve maintenance planning, condition assessment, and remaining-life estimation.

Manufacturing became another major application domain. Rosen et al. (2015) showed how DT concepts could support more autonomous production systems. Schleich et al. (2017), Uhlemann, Lehmann, and Steinhilper (2017), Söderberg et al. (2017), and Leng et al. (2019) further demonstrated how DT could improve shop floor transparency, geometric control, system monitoring, and smart workshop management. Aivaliotis et al. (2019) extended this practical value into predictive maintenance and remaining useful life estimation, while Platenius-Mohr et al. (2020) highlighted interoperability through Asset Administration Shell-based transformation. These applications show that manufacturing has provided one of the cleanest environments for the operationalisation of DT, especially where productivity, accuracy, and predictive capability are essential.

Over time, DT has also expanded into product lifecycle management, robotics, and systems engineering. Grieves and Vickers (2016) connected DT directly to lifecycle decision-making, while Schluse and Rossmann (2016) demonstrated experimentable DTs in engineering and robotics contexts. Canedo (2016) and Alam and El Saddik (2017) contributed from the informatics and CPS perspective by showing how DT could support lifecycle optimisation and system integration. Tao, Qi, et al. (2019) further strengthened this position by showing how DT could support design, manufacture, and service integration through big data and lifecycle intelligence. These contributions indicate that DT is not merely a visual model but a broader digital strategy for connecting processes, information, and decisions over time.

More recent applications show that DT has expanded well beyond its early industrial roots. In healthcare, Hassani, Huang and MacFeely (2022) explored sectoral use of DT, while Madubuike and Anumba (2022) demonstrated DT-aided healthcare facility management through the Shanghai Tongji Hospital case. In energy systems, Kandasamy et al. (2022) applied DT to a smart-grid cybersecurity testbed, while Mrzyk et al. (2023) and Pacheco-Blazquez et al. (2024) showed its use in predictive maintenance and remaining useful life assessment in wind-related energy infrastructure. In rail and transport infrastructure, Doubell (2023) applied DT in the railway context, while the table also shows its relevance to local road network monitoring and infrastructure management. These examples illustrate that DT is increasingly used wherever systems are operationally complex, data-intensive, and maintenance-sensitive.

The built environment is also becoming more visible within the documented application of DT. Borth, Verriet and Muller (2019) applied DT/CPS concepts to architectural design and development workflows, while Chacón et al. (2024) demonstrated DT use in the monitoring of office building construction processes. The table also shows more recent building-related applications at campus and office scale, including operational integration and interactive building management. These examples are important because they suggest that the built environment is beginning to shift from conceptual interest to more practical, demonstrable DT use. This is particularly significant for building operations and indoor environmental quality (IEQ) management, where real-time monitoring, integrated data, and responsive control are becoming increasingly important.

A consistent pattern across the documented applications is the strong role of predictive maintenance, operational visibility, and informed decision-making. In many industries, DT is valued for enabling users to detect problems early, interpret large volumes of data, test responses digitally, and improve performance before failure occurs. This is why predictive maintenance remains one of the strongest recurring themes in the academic record (Aivaliotis et al., 2019). At the same time, the table shows that DT also supports broader purposes such as lifecycle management, safety, interoperability, cybersecurity testing, building operations, and strategic control.

The documented practical use of DT within the existing body of knowledge shows a clear movement from conceptual development to real-world application. Table 3.1 confirms that DT is now part of a wider interdisciplinary and intersectoral body of knowledge. Its growing use across aerospace, manufacturing, healthcare, energy, transport, infrastructure, and the built environment shows that it has become a practical digital strategy for managing complex systems more intelligently. In this regard, the documented evidence provides a strong foundation for understanding why DT is increasingly relevant to the built environment and, more specifically, to IEQ management.

3.8 Summary

This chapter explored the application of DT across a range of important economic industries and showed that the technology has moved far beyond its original industrial roots. It was established that DT is increasingly used as a strategic tool to improve performance, efficiency, reliability, safety, and decision-making in complex operational environments. Across the industries discussed, the chapter showed that DT supports the creation of virtual representations of physical assets, systems, and processes, allowing organisations to monitor performance, simulate conditions, predict failures, and optimise operations more effectively.

The chapter examined the use of DT in the aerospace, automotive, manufacturing, oil and gas, mining, energy, and healthcare industries. In each case, the discussion showed that the technology is being applied in ways that address the sector's specific needs, whether through design improvements, predictive maintenance, production optimisation, safety enhancements, equipment monitoring, or personalised service delivery. The chapter also highlighted that DT is relevant to other sectors, such as the built environment, infrastructure, facility management, and smart city systems, thereby demonstrating its wide applicability across the economy.

At the same time, the chapter acknowledged that the implementation of DT is not without challenges. Issues such as data integration, cybersecurity, infrastructure demands, technical complexity, cost, and regulatory concerns continue to affect deployment in different sectors. Even so, the chapter made it clear that the benefits of DT remain substantial. Ultimately, the chapter showed that DT is becoming an increasingly important technology for economic industries seeking smarter, more proactive, and more data-driven ways to operate and manage complex systems.

References

Ahmed, H. and Devoto, L. (2020) 'The potential of a Digital Twin in surgery', *Artificial Cells, Nanomedicine, and Biotechnology*, 28(4), pp. 509–510. doi:10.1177/1553350620975896

Aivaliotis, P. et al. (2019) 'Methodology for enabling digital twin using advanced physics-based modelling in predictive maintenance', in *Proceedings of the 52nd CIRP Conference on Manufacturing Systems*. Elsevier, pp. 417–422. doi:10.1016/j.procir.2019.03.072

Alam, K.M. and El Saddik, A. (2017) 'C2PS: A digital twin architecture reference model for the cloud-based cyber-physical systems', *IEEE Access*, 5, pp. 2050–2062. doi:10.1109/ACCESS.2017.2657006

Arsecularatne, B., Rodrigo, N. and Chang, R. (2024) 'Digital Twins for reducing energy consumption in buildings: A review', *Sustainability*, 16(21), p. 9275. doi:10.3390/su16219275

Ashrafian, A. and Pedersen, S. (2023) 'Digital Twin for complex logistics systems: The case study of a large-scale automated order picking and fulfillment system', *IFAC-PapersOnLine*, 56(2), pp. 11056–11061. doi:10.1016/j.ifacol.2023.10.808

Attaran, M. and Celik, B.G. (2023) 'Digital Twin: Benefits, use cases, challenges, and opportunities', *Decision Analytics Journal*, 6, p. 100165. doi:10.1016/J.DAJOUR.2023.100165

Augustine, P. (2020) 'The industry use cases for the Digital Twin idea', *Advances in Computers*, 117(1), pp. 79–105. doi:10.1016/BS.ADCOM.2019.10.008

Bende, A. and Gochhait, S. (2020) 'Leveraging Digital Twin technology in the healthcare industry – A machine learning based approach', *European Journal of Molecular & Clinical Medicine*, 7(6), pp. 2547–2557. Available at: https://www.psychosocial.com/index.php/ijpr/article/view/2388

Bhatti, G., Mohan, H. and Raja Singh, R. (2021) 'Towards the future of smart electric vehicles: Digital twin technology', *Renewable and Sustainable Energy Reviews*, 141, p. 110801. doi:10.1016/J.RSER.2021.110801

Borth, M., Verriet, J. and Muller, G. (2019) 'Digital twin strategies for SoS: 4 challenges and 4 architecture setups for digital twins of SoS', in *Proceedings of the 14th Annual Conference System of Systems Engineering (SoSE)*, pp. 164–169. doi:10.1109/SYSOSE.2019.8753860

Canedo, A. (2016) 'Industrial IoT lifecycle via digital twins', in *Proceedings of the International Conference on Hardware/Software Codesign and System Synthesis (CODES+ISSS)*. doi:10.1145/2968456.2974007

Chacón, R. et al. (2024) 'Digital twinning of building construction processes: Case study of a reinforced concrete cast-in structure', *Journal of Building Engineering*, 84, p. 108522. doi:10.1016/j.jobe.2024.108522

Chen, L. et al. (2021) 'Parallel mining operating systems: From digital twins to mining intelligence', in *Proceedings of the IEEE 1st International Conference on Digital Twins and Parallel Intelligence (DTPI)*, pp. 469–473. doi:10.1109/DTPI52967.2021.9540195

Damjanovic-Behrendt, V. (2018) 'A Digital Twin-based privacy enhancement mechanism for the automotive industry', in *Proceedings of the 9th International Conference on Intelligent Systems on Theory, Research and Innovation in Applications*, pp. 272–279. doi:10.1109/IS.2018.8710526

Doubell, G.C., Basson, A.H., Kruger, K. and Conradie, P.D.F. (2023) 'A Digital Twin system for railway infrastructure', *R&D Journal*, 39, pp. 23–34. doi:10.17159/2309-8988/2023/v39a3

El Bazi, N. et al. (2023) 'Generic Multi-layered Digital-Twin-framework-enabled asset lifecycle management for the sustainable mining industry', *Sustainability*, 15(4), p. 3470. doi:10.3390/SU15043470

Erol, T., Mendi, A.F. and Dogan, D. (2020) 'The Digital Twin revolution in healthcare', in *Proceedings of the 4th International Symposium on Multidisciplinary Studies and Innovative Technologies (ISMSIT)*. doi:10.1109/ISMSIT50672.2020.9255249

Feldt, J., Kourouklis, T., Kontny, H. and Wagenitz, A. (2020) 'Digital twin: revealing potentials of real-time autonomous decisions at a manufacturing company', *Procedia CIRP*, 88, pp. 185–190. doi:10.1016/j.procir.2020.05.033

Fotland, G., Haskins, C. and Rølvåg, T. (2020) 'Trade study to select best alternative for cable and pulley simulation for cranes on offshore vessels', *Systems Engineering*, 23(2), pp. 177–188. doi:10.1002/SYS.21503

Glaessgen, E.H. and Stargel, D.S. (2012) 'The digital twin paradigm for future NASA and U.S. Air force vehicles', in *Proceedings of the 53rd Structures, Structural Dynamics, and Materials Conference*. American Institute of Aeronautics and Astronautics, pp. 1–14. doi:10.2514/6.2012-1818

Grieves, M. and Vickers, J. (2016) 'Digital twin: Mitigating unpredictable, undesirable emergent behavior in complex systems', in F.-J. Kahlen, S. Flumerfelt and A. Alves (eds.), *Transdisciplinary perspectives on complex systems: New findings and approaches*. Cham, Switzerland: Springer, pp. 85–113. doi:10.1007/978-3-319-38756-7_4

Hassani, H., Huang, X. and MacFeely, S. (2022) 'Impactful Digital Twin in the healthcare revolution', *Big Data and Cognitive Computing*, 6(3), p. 83. doi:10.3390/BDCC6030083

Hwang, J., Kim, J. and Yoon, S. (2025) 'DT-BEMS: Digital twin-enabled building energy management system for information fusion and energy efficiency', *Energy*, 326, p. 136162. doi:10.1016/j.energy.2025.136162

Jimenez, J.I., Jahankhani, H. and Kendzierskyj, S. (2020) 'Health care in the cyberspace: Medical cyber-physical system and digital twin challenges', in M. Farsi, A. Daneshkhah, A. Hosseinian-Far and H. Jahankhani (eds.), *Digital Twin technologies and smart cities*. Cham, Switzerland: Springer, pp. 79–92. doi:10.1007/978-3-030-18732-3_6

Kan, C. and Anumba, C.J. (2019) 'Digital Twins as the next phase of cyber-physical systems in construction', in *Proceedings of the ASCE International Conference on Computing in Civil Engineering*. Atlanta: American Society of Civil Engineers (ASCE), pp. 256–264. doi:10.1061/9780784482438.033

Kandasamy, N.K., Venugopalan, S. and Wong, T.K. (2022) 'An electric power digital twin for cyber security testing, research and education', *Computers & Electrical Engineering*, 101, p. 108061. doi:10.1016/j.compeleceng.2022.108061

Kraft, E.M. (2016) 'The US air force digital thread/Digital Twin – Life cycle integration and use of computational and experimental knowledge', in *Proceedings of the 54th AIAA Aerospace Sciences Meeting*. doi:10.2514/6.2016-0897

Kritzinger, W. et al. (2018) 'Digital Twin in manufacturing: A categorical literature review and classification', *IFAC-PapersOnLine*, 51(11), pp. 1016–1022. doi:10.1016/j.ifacol.2018.08.474

Kunath, M. and Winkler, H. (2018) 'Integrating the Digital Twin of the manufacturing system into a decision support system for improving the order management process', in *Proceedings of the 51st CIRP Conference on Manufacturing Systems*. Elsevier, pp. 225–231. doi:10.1016/j.procir.2018.03.192

Lasi, H. et al. (2014) 'Industry 4.0', *Business and Information Systems Engineering*, 6(4), pp. 239–242. doi:10.1007/S12599-014-0334-4

Leng, J. et al. (2019) 'Digital twin-driven manufacturing cyber-physical system for parallel controlling of smart workshop', *Journal of Ambient Intelligence and Humanized Computing*, 10(3), pp. 1155–1166. doi:10.1007/S12652-018-0881-5

Li, B.H. et al. (2017) 'Simulation-based cyber-physical systems and Internet-of-Things', in S. Mittal, U. Durak and T. Ören (eds.), *Guide to simulation-based disciplines*. Cham: Springer, Switzerland pp. 103–126. doi:10.1007/978-3-319-61264-5_5

Li, L. et al. (2022) 'Digital Twin in aerospace industry: A gentle introduction', *IEEE Access*, 10, pp. 9543–9562. doi:10.1109/ACCESS.2021.3136458

Lu, Y. et al. (2020) 'Digital Twin-driven smart manufacturing: Connotation, reference model, applications and research issues', *Robotics and Computer-Integrated Manufacturing*, p. 101837. doi:10.1016/j.rcim.2019.101837

Ma, T. et al. (2025) 'Digital twin for 3D interactive building operations: Integrating BIM, IoT-enabled building automation systems, AI, and mixed reality', *Automation in Construction*, 176, p. 106277. doi:10.1016/j.autcon.2025.106277

Machalski, A. et al. (2025) 'The concept of a Digital Twin for the Wały Śląskie hydroelectric power plant: A case study in Poland', *Energies*, 18(8), p. 2021. doi:10.3390/en18082021

Madubuike, O.C. and Anumba, C.J. (2022) 'Digital Twin application in healthcare facilities management', in R.R.A. Issa (ed.), *Computing in civil engineering 2021*. Reston, VA: American Society of Civil Engineers, pp. 395–402. doi:10.1061/9780784483893.046

Marmolejo-Saucedo, J.A. (2020) 'Design and development of Digital Twins: A case study in supply chains', *Mobile Networks and Applications*, 25(6), pp. 2141–2160. doi:10.1007/s11036-020-01557-9

Mendi, A.F. (2022) 'A Digital Twin case study on automotive production line', *Sensors*, 22(18), p. 6963. doi:10.3390/s22186963

Mrzyk, P. et al. (2023) 'Digital Twins for predictive maintenance: A case study for a flexible IT-architecture', *Procedia CIRP*, 119, pp. 152–157. doi:10.1016/j.procir.2023.03.087

Nasim, M., Rajabifard, A., Chen, Y. and Samali, B. (2025) 'A demonstration of a digital twin framework for structural health monitoring: Application to bridge infrastructures', *Journal of Infrastructure Intelligence and Resilience*, 5(1), p. 100184. doi:10.1016/j.iintel.2025.100184

Negri, E., Fumagalli, L. and Macchi, M. (2017) 'A review of the roles of Digital Twin in CPS-based production systems', *Procedia Manufacturing*, 11, pp. 939–948. doi:10.1016/j.promfg.2017.07.198

Ngwenya, K.M., Issa, R.R.A. and Anumba, C. (2021) 'Visualization of cost funding sources in building information models', in R.R.A. Issa (ed.), *Computing in civil engineering 2021*. Reston, VA: American Society of Civil Engineers, pp. 1376–1383. doi:10.1061/9780784483893.168

Pacheco-Blazquez, R. et al. (2024) 'A Digital Twin for assessing the remaining useful life of offshore wind turbine structures', *Journal of Marine Science and Engineering*, 12(4), p. 573. doi:10.3390/jmse12040573

Park, J.-Y., Lee, S.-H. and Kim, J.-H. (2025) 'Digital-Twin tool for a drinking water distribution system using augmented reality and EPANET', *Environmental Modelling & Software*, 184. doi:10.1016/j.envsoft.2025.106829

Phanden, R.K., Sharma, P. and Dubey, A. (2021) 'A review on simulation in digital twin for aerospace, manufacturing and robotics', *Materials Today: Proceedings*, 38, pp. 174–178. doi:10.1016/J.MATPR.2020.06.446

Piromalis, D. and Kantaros, A. (2022) 'Digital Twins in the automotive industry: The road toward physical-digital convergence', *Applied System Innovation*, 5(4), p. 65. doi:10.3390/ASI5040065

Platenius-Mohr, M., Malakuti, S., Grüner, S., Schmitt, J. and Goldschmidt, T. (2020) 'File- and API-based interoperability of digital twins by model transformation: An IIoT case study using asset administration shell', *Future Generation Computer Systems*, 113, pp. 94–105. doi:10.1016/J.FUTURE.2020.07.004

Qi, Q. et al. (2018) 'Digital Twin service towards smart manufacturing', *Procedia CIRP*, 72, pp. 237–242. doi:10.1016/j.procir.2018.03.103

Rabah, S. et al. (2018) 'Towards improving the future of manufacturing through digital twin and augmented reality technologies', in *Proceedings of the 28th International Conference on Flexible Automation and Intelligent Manufacturing (FAIM)*, pp. 460–467. doi:10.1016/j.promfg.2018.10.070

Rosen, R. et al. (2015) 'About the importance of autonomy and Digital Twins for the future of manufacturing', *IFAC-PapersOnLine*, pp. 567–572. doi:10.1016/j.ifacol.2015.06.141

Schleich, B. et al. (2017) 'Shaping the digital twin for design and production engineering', *CIRP Annals - Manufacturing Technology*, 66(1), pp. 141–144. doi:10.1016/j.cirp.2017.04.040

Schluse, M. and Rossmann, J. (2016) 'From simulation to experientable digital twins: Simulation-based development and operation of complex technical systems', in *Proceedings of the IEEE International Symposium on Systems Engineering (ISSE)*, pp. 1–6. doi:10.1109/SysEng.2016.7753162

Shahzad, M. et al. (2022) 'Digital Twins in built environments: An investigation of the characteristics, applications, and challenges', *Buildings*, 12(2), p. 120. doi:10.3390/BUILDINGS12020120

Singh, M. et al. (2022) 'Applications of Digital Twin across Industries: A review', *Applied Sciences*, 12(11), p. 5727. doi:10.3390/APP12115727

Sleiti, A.K., Kapat, J.S. and Vesely, L. (2022) 'Digital twin in energy industry: Proposed robust digital twin for power plant and other complex capital-intensive large engineering systems', *Energy Reports*, 8, pp. 3704–3726. doi:10.1016/J.EGYR.2022.02.305

Söderberg, R. et al. (2017) 'Toward a Digital Twin for real-time geometry assurance in individualized production', *CIRP Annals - Manufacturing Technology*, 66(1), pp. 137–140. doi:10.1016/j.cirp.2017.04.038

Steyn, W.J.M. and Broekman, A. (2022) 'Development of a Digital Twin of a local road network: A case study', *Journal of Testing and Evaluation*, 50(6), pp. 2901–2915. doi:10.1520/JTE20210043

Tao, F. and Qi, Q. (2019) 'Make more digital twins', *Nature*. doi:10.1038/d41586-019-02849-1

Tao, F., Qi, Q., et al. (2019) 'Digital Twins and cyber–physical systems toward smart manufacturing and Industry 4.0: Correlation and comparison', *Engineering*, 5(4), pp. 653–661. doi:10.1016/j.eng.2019.01.014

Tao, F., Zhang, M. and Nee, A.Y.C. (2019a) 'Applications of Digital Twin', in B. Guerin (ed.), *Digital Twin driven smart manufacturing*. London: Academic Press, pp. 29–62. doi:10.1016/B978-0-12-817630-6.00002-3

Tuegel, E.J. et al. (2011) 'Reengineering aircraft structural life prediction using a digital twin', *International Journal of Aerospace Engineering*, 2011, pp. 1–14. doi:10.1155/2011/154798

Uhlemann, T.H.J., Lehmann, C. and Steinhilper, R. (2017) 'The Digital Twin: Realizing the cyber-physical production system for Industry 4.0', in *Proceedings of the 24th CIRP Conference on Life Cycle Engineering*, pp. 335–340. doi:10.1016/j.procir.2016.11.152

Wanasinghe, T.R. et al. (2020) 'Digital Twin for the oil and gas industry: Overview, research trends, opportunities, and challenges', *IEEE Access*, 8, pp. 104175–104197. doi:10.1109/ACCESS.2020.2998723

Wang, T. et al. (2024) 'Pipeline condition monitoring towards digital twin system: A case study', *Journal of Manufacturing Systems*, 73, pp. 256–274. doi:10.1016/j.jmsy.2024.02.006

Xiong, M. and Wang, H. (2022) 'Digital twin applications in aviation industry: A review', *International Journal of Advanced Manufacturing Technology*, 121(9–10), pp. 5677–5692. doi:10.1007/S00170-022-09717-9

Yu, W. et al. (2022) 'Energy digital twin technology for industrial energy management: Classification, challenges and future', *Renewable and Sustainable Energy Reviews*, 161, p. 112407. doi:10.1016/J.RSER.2022.112407

Zhang, C. et al. (2019) 'A reconfigurable modeling approach for digital twin-based manufacturing system', in *Proceedings of the 11th CIRP Conference on Industrial Product-Service Systems*. Elsevier, pp. 118–125. doi:10.1016/j.procir.2019.03.141

4 Digital Twin for Built Assets

4.1 Introduction

The built environment is undergoing a gradual but important digital transformation. Across the world, buildings and infrastructure are increasingly being viewed not only as physical assets but also as intelligent systems that can generate, receive, and respond to information. This shift has created new opportunities for the design, construction, operation, maintenance, and management of built assets throughout their lifecycle. Within this transformation, Digital Twin (DT) has emerged as one of the most promising technologies for strengthening the relationship between physical buildings and their digital representations. By linking real-world assets with dynamic virtual models, DT provides a more advanced basis for monitoring, analysis, simulation, and decision-making in the built environment (Khajavi et al., 2019; Opoku et al., 2021; Shahzad et al., 2022).

The relevance of DT to the built environment is closely connected to the complexity of buildings and infrastructure. Unlike isolated industrial products, built assets are long-lived systems comprising interconnected structural, mechanical, electrical, environmental, and operational components. They are also influenced by occupancy behaviour, maintenance practices, environmental conditions, and management decisions over time. As a result, effective building management requires more than static records or fragmented digital tools. It requires systems that can integrate information from multiple sources, support real-time monitoring, and provide intelligent feedback on current and future performance (Nie et al., 2019; Madubuike and Anumba, 2022). DT responds to this need by creating a digital environment in which the condition and behaviour of a building can be understood more holistically.

The development of DT within the built environment did not occur in isolation. It builds on earlier advances in digitalisation, including computer-aided design (CAD), Building Information Modelling (BIM), sensor technologies, data analytics, and cyber-physical systems (CPS). BIM, in particular, provided an important foundation by enabling the structured digital modelling of buildings and their components. However, while BIM supports information-rich representation and collaboration, DT takes this progression further

DOI: 10.1201/9781003743330-4

by introducing real-time data exchange, continuous updating, simulation, and operational intelligence (Ito, 2019; Kan and Anumba, 2019). In this way, DT is increasingly being viewed as a more advanced and responsive digital system for the built environment.

The use of DT in built assets has notable impacts on design, construction, operation, facility management, and lifecycle control. It can improve building efficiency, strengthen maintenance planning, enhance stakeholder coordination, support predictive insights, and contribute to smarter, more sustainable asset management (Alonso et al., 2019; BSI Group, 2019; Hosamo et al., 2023). At the same time, its implementation also raises important questions about data integration, interoperability, complexity, cost, and digital readiness within the industry.

This chapter explores DT within the built environment by examining its role in the digital transformation of built assets. It discusses how DT can be adapted for building use, its relevance to data management, real-time monitoring, decision-making, building efficiency, facility management, and lifecycle management. It also considers the progression from BIM to DT, as well as the benefits and challenges associated with implementing DT for built assets. In doing so, the chapter provides a foundation for understanding DT as a transformative technology for the future of the built environment.

4.2 Digital Transformation for Built Assets

The advancement of technology continues to drive innovation, which remains a critical factor in improving performance within the building industry. Across the world, rising benchmarks for innovation are reshaping how buildings are designed, delivered, and managed. Yet, the execution of certain operational activities within the industry has remained constrained by slow adoption and resistance to change (Aghimien, Aigbavboa and Oke, 2018). Dithebe et al. (2018) contend that many employers in the building industry remain reluctant to approve new and improved building techniques and more creative approaches that could contribute positively and efficiently to industry improvement. This hesitation may arise from concerns about increased cost, uncertainty about untested methods, fear of delays, and limited familiarity with newer technologies. In some cases, the lack of awareness of long-term benefits also contributes to this resistance. This situation highlights the importance of effective communication, stakeholder engagement, and industry education in encouraging broader acceptance of innovation (Alaloul et al., 2020). Alaloul et al. (2020) further emphasised that digitalisation can benefit the building industry by substantially improving productivity through the implementation of advanced building techniques and processes.

The building industry is also facing persistent challenges such as skills shortages and declining productivity. Hasan (2015) argued that digitalisation can help to address these gaps. However, the growing presence of digital tools has also created anxiety within parts of the industry, particularly around the fear that digitalisation may replace human workers with machines and robots. Even so, professionals within the building industry are among the least likely to be fully replaced by machines, because digitalisation does not replicate the full range of human judgement, creativity, and contextual intelligence required in the built environment (Dall'Omo, 2017). Oke et al. (2018) reiterated that the building industry is widely known for clinging to established practices and resisting change, even as rapid digital transformation continues worldwide. This makes the digital transformation of built assets not only a technical issue but also an organisational and cultural one.

Digitisation can, however, automate monotonous and repetitive tasks to improve efficiency, increase profitability, and enable faster service delivery. Over recent decades, numerous software packages and digital systems have been developed for use by built environment professionals (Royal Institution of Chartered Surveyors [RICS, 2017]). These technologies support a more digital approach to professional tasks and reduce dependence on slower, more traditional methods of working. They also create opportunities to work more efficiently and more intelligently. More importantly, they enable the capture and analysis of large volumes of live data, which in turn helps professionals make more informed decisions and achieve better project outcomes (Aghimien et al., 2018). RICS (2017) states that digital services will have a disruptive and significant impact on the industry. This impact is not unique to the building industry. In many professional fields, once stakeholders become aware of available technologies, they expect professionals to deliver more value by leveraging the benefits those tools offer. At the same time, history has shown that the mechanisation of one task does not mean the profession itself disappears. Rather, the traditional role of building professionals evolves in response to technological change, but it does not become obsolete (Aghimien et al., 2018).

The construction industry has therefore undergone a major transformation, moving from more traditional ways of building towards digitalisation (Akinshipe, Aliu and Aigbavboa, 2022). A range of digital tools and systems is driving this transformation. These include augmented reality (AR), simulation technologies, big data and analytics, the Internet of Things (IoT), autonomous robots, cybersecurity systems, cloud technologies, additive manufacturing, and horizontal and vertical system integration (Jazzar et al., 2020). Each of these technologies contributes in different ways to the digital transformation of built assets and to the broader improvement of industry processes.

BIM, for example, is a virtual modelling technology that supports planning, design, construction, and facility management activities (Jung, 2017). CAD has long been used in architectural and construction practice to improve the speed and efficiency of drawing production compared with traditional paper-based methods. Virtual reality and AR allow users not only to view digital information but also to interact with content and collaborate with others in real time, even when working remotely (Omer et al., 2019). Adaptive robotics and artificial intelligence are increasingly used in manufacturing-related activities such as design, production, and assembly processes (Aghimien et al.,

2022). Additive manufacturing and three-dimensional (3D) printing enable the creation of complex physical objects directly from CAD models. These processes can reduce waste, accommodate different materials, produce more complex forms, and lower construction costs (Knapp et al., 2017). Collectively, these digital tools are reshaping the construction industry by improving efficiency, reducing waste, increasing cost-effectiveness, and supporting more integrated ways of working (Tanga et al., 2021; Adeniran and Shakantu, 2023).

The digital transformation of built assets reflects a broader movement towards more intelligent, connected, and data-driven ways of delivering and managing the built environment. It is changing how professionals work, how decisions are made, and how assets are designed, constructed, and operated. Although resistance, skills gaps, and adoption barriers still exist, the direction of change is clear. Digitalisation is becoming increasingly central to the future of built assets and to the long-term transformation of the building industry.

4.3 Adapting Digital Twin for Built Environment Use

So far, DT has been utilised across a wide range of industries. The aerospace industry was the pioneer sector to embrace and apply the technique. With the rise of smart manufacturing, manufacturing also began to use DT to improve efficiency, productivity, and operational performance (Kan and Anumba, 2019). The successful use of DT in these industries has laid a strong foundation for its adoption in the built environment. It has been shown that DT can extend beyond conventional industrial systems and be applied to other complex, data-intensive settings where performance, coordination, and lifecycle management are important.

The built environment is a particularly suitable context for DT because buildings and infrastructure are not static. They are dynamic assets that change over time through occupation, operation, maintenance, repair, refurbishment, and adaptation. They also comprise multiple interconnected systems, including structural elements, building services, environmental systems, control systems, and user-driven activities. For this reason, the built environment requires a technology that can support continuous monitoring, integrated information flow, predictive insight, and more informed decision-making throughout the asset lifecycle. DT provides such a platform by linking the physical asset to a virtual representation that can receive, process, and interpret real-time and historical data (Khajavi et al., 2019; Opoku et al., 2021; Shahzad et al., 2022).

The adaptation of DT for use in the built environment is therefore not simply a transfer of technology from one industry to another. It is a purposeful adjustment of the DT concept to suit the unique characteristics of buildings and infrastructure. Unlike many manufactured products, built assets are long-lived systems that are spatially fixed, operationally complex, and managed by multiple stakeholders over long periods. Their performance depends not only on technical components but also on occupancy patterns, environmental conditions, maintenance quality, and management decisions.

This means that DT in the built environment must be capable of doing more than creating a digital representation. It must also support building performance monitoring, information integration, operational control, facility management, and lifecycle planning (Nie et al., 2019; Madubuike and Anumba, 2022; Hosamo et al., 2023).

Figure 4.1 illustrates this adaptability of DT for built environment use. As shown in the figure, DT can be used for real-time monitoring in several important ways. These include service life prediction, ensuring product reliability, quality optimisation, improved maintenance, and data capturing and integration. These functions are especially important in buildings because physical assets and services often deteriorate gradually. Through connected sensors, building systems, and operational data streams, DT can help managers and operators monitor the condition of building components and predict their likely performance over time. This supports more proactive maintenance, better reliability, and improved service delivery within buildings and facilities (Khajavi et al., 2019; Nie et al., 2019; Hosamo et al., 2023). The ability to capture and integrate data is equally important because built assets generate data from multiple sources, including sensors, meters, control systems, inspection records, and user feedback. DT helps bring these together into one coordinated environment where information can be interpreted more effectively (Opoku et al., 2021; Shahzad et al., 2022).

Figure 4.1 also shows that DT can provide important analytical capabilities for the built environment. These include predicting future performance, managing the IoT lifecycle, supporting decision-making, and promoting future planning. These analytical functions enhance the value of DT by moving it beyond simple observation. In the built environment, decision-makers often need to assess how current conditions may affect future performance, costs, user comfort, maintenance demands, and system reliability. DT can support this by analysing historical and live data, simulating future scenarios, and helping stakeholders make better decisions regarding operations, interventions, and long-term planning (Kan and Anumba, 2019; Re Cecconi et al., 2021;

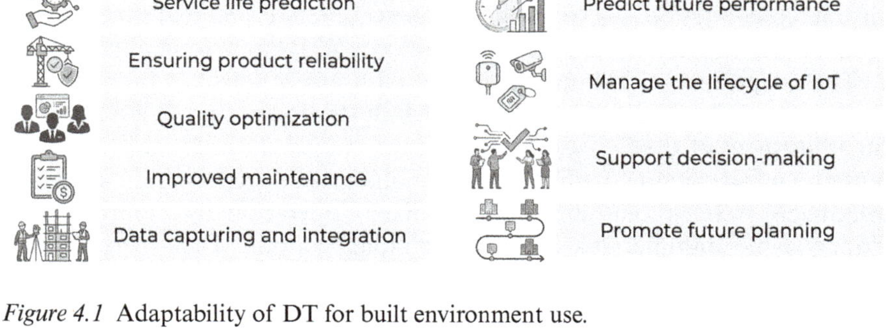

Figure 4.1 Adaptability of DT for built environment use.

Hosamo et al., 2023). In smart and connected buildings, where IoT devices are increasingly used for sensing and control, DT can also support the management of these connected systems across their lifecycle by linking device performance with broader building objectives (Nie et al., 2019; Shahzad et al., 2022).

Another important reason why DT is adaptable to the built environment is its support for collaboration among stakeholders. Buildings are planned, delivered, operated, and maintained by many actors, including designers, contractors, owners, facility managers, service providers, and users. DT can serve as a common digital environment through which these stakeholders access more coordinated information and make better informed decisions about the asset (Araszkiewicz, 2017; Dawkins, Hudson-Smith and Dennett, 2018; Madubuike and Anumba, 2022). This makes DT valuable not only for technical monitoring but also for management, communication, and lifecycle governance.

The prior use of DT in other industries has paved the way for the building industry to recognise its possibilities. These earlier applications have shown that DT can improve performance, support predictive insight, strengthen maintenance, and guide more intelligent planning. In the built environment, these capabilities translate into improved monitoring, integrated data management, stronger decision-making, and more effective lifecycle control. The adaptation of DT for built environment use, therefore, represents an important step in the digital transformation of built assets and in the development of smarter, more responsive, and better managed buildings and infrastructure (Kan and Anumba, 2019; Khajavi et al., 2019; Opoku et al., 2021).

4.3.1 Data Management

DT relies heavily on large volumes of data in order to function properly. Rosen et al. (2015) established that implementing DT requires substantial digital data storage capacity. Schleich et al. (2017) also confirmed that big data management and analytics have become major concerns in the DT context, particularly regarding model scalability, interoperability, expansibility, and fidelity. In the DT environment, managing large, complex datasets is not a secondary issue. It is one of the core conditions for the system's successful functioning. This is because DT depends on the continuous capture, storage, processing, exchange, and interpretation of data from physical and virtual environments. Without effective data management, the DT cannot accurately reflect the condition, behaviour, or performance of the physical asset.

Tao et al. (2018) further explored the use of large datasets, referred to as Big Data, as a tool for product lifecycle optimisation through DT. This position is particularly relevant to the built environment, where data are generated throughout the entire project and asset lifecycle. The construction industry is increasingly being pushed into the era of big data by rapid global technological advancement. As a result, the importance of data mining and data analysis has become more prominent, especially when dealing with the large volumes of data created across the project lifecycle, including conceptualisation, planning, design, construction, operation, and management (Tchana, Ducellier and Remy, 2019a). In the built environment, this data may come from design models, construction records, sensor networks, maintenance histories, inspection reports, operational systems, and user interactions. The challenge is therefore not only the volume of data, but also its variety, speed, and long-term relevance.

As Big Data applications continue to expand in the building industry, interoperability has become a critical issue. Data generated across the lifecycle of built assets often originates from different systems, stakeholders, software environments, and technical platforms. This creates difficulties in integration, consistency, and exchange. For this reason, the need for open data standards and a uniform exchange format has become a crucial issue that must be widely recognised within the industry (Nie et al., 2019). If data cannot move effectively between systems, then the value of DT becomes limited, regardless of how advanced the model itself may be.

DT can help address this challenge by supporting active interoperability, open data standards, and more uniform data exchange. This is possible because DT promotes close synchronisation between the virtual and physical components of a project or asset (Khajavi et al., 2019). Through this synchronisation, data from different sources can be linked more meaningfully, thereby creating a more coordinated and reliable digital environment. A further advantage is that DT allows direct comparison between theoretical data values and actual project conditions. This means that projected performance, design assumptions, and planned outcomes can be continuously assessed against real operational behaviour.

With DT, the entire project and asset lifecycle can be simulated, visualised, assessed, tracked, evaluated, reviewed, enhanced, and validated on the same platform by multiple stakeholders (Wang et al., 2020). This creates a stronger basis for collaborative decision-making and more effective lifecycle management. In addition, an open-access and well-structured data platform can support standardised processes, workflows, and procedures across projects. Such a platform would benefit the building industry by promoting more effective, transparent, and collaborative ways of working (Kunz and Fischer, 2020). In this sense, data management is not merely a technical support function within DT. It is one of the key foundations upon which successful DT adaptation for built-environment use relies.

4.3.2 Real-Time Monitoring

One of the key functional streams of DT is its ability to monitor in real time. This is one of the features that gives DT much of its practical value. In many applications, studies have focused either on monitoring the product itself or on monitoring the process through which it functions. Real-time monitoring is therefore widely recognised as a core function of DT because it allows the condition, behaviour, and performance of a physical system to be observed continuously through its virtual counterpart (Dawkins, Hudson-Smith and Dennett,

2018). This continuous visibility makes it possible to identify changes as they occur and to respond more quickly and more intelligently than would be possible through periodic inspection alone.

Evidence of this can be seen in other industries where DT has already been applied successfully. In the work of Glaessgen and Stargel (2012), simulation was used alongside an existing management system to continuously monitor the health of a spacecraft. In a similar way, Li et al. (2017) used DT to track the state of aircraft damage and to predict possible crack initiation and propagation. Söderberg et al. (2017) also used real-time monitoring in individualised manufacturing in order to ensure geometric accuracy and precision throughout planning, design, and production. These examples show that DT is valuable not only for representing a system but also for monitoring its condition in a live, operational manner. In the built environment, this monitoring capability is especially important because buildings, infrastructure, and construction activities are affected by changing physical, environmental, and human conditions. Built assets are exposed to wear, load, environmental stress, usage patterns, and operational changes over time. Real-time monitoring through DT therefore provides an opportunity to continuously observe these changes, compare them with expected conditions, and detect emerging problems before they become severe. This is particularly important for safety, maintenance, quality control, and operational efficiency.

Another important area in which DT is being used is labour behaviour monitoring, especially to reduce human error and improve safety. Grieves and Vickers (2016) sought to reduce unpredictable, unwanted emergent behaviour in complex systems to mitigate forms of human engagement that could lead to mishaps. This is highly relevant to construction because the industry has long been characterised by poor safety performance. As a labour-intensive industry, construction is often affected by human error, unsafe behaviour, and rapidly changing site conditions. These factors continue to contribute to safety problems on site. DT provides an avenue for reducing such risks by monitoring both site conditions and worker behaviour in real time.

In the construction context, a DT algorithm can be used in at least two important ways. First, it can be used to study dynamic site conditions as they change during construction. Second, it can be used to assess the safety condition of a facility or operation by comparing actual site conditions against an established set of standards or acceptable conditions (Zhou et al., 2019). This makes DT useful not only for technical system monitoring but also for real-time safety supervision and control. For example, the imaging systems and algorithms associated with DT enable the early detection of cracks in concrete structures and other physical defects. In such cases, accidents may be prevented by triggering further inspections and safety checks before the issue becomes critical (Kan and Anumba, 2019). More importantly, DT supports real-time monitoring and reconstruction of site activities, helping prevent workers from engaging in dangerous behaviour. Alert systems linked to DT can be used to notify managers, safety officers, and workers who may be exposed to risk on site (Osunsanmi, Oke and Aigbavboa, 2018).

Real-time monitoring is one of the most important ways DT can be adapted for use in the built environment. It strengthens safety, improves quality control, supports early fault detection, and enables more proactive decision-making across building and construction activities. For built assets that require continuous oversight and timely intervention, this capability is especially valuable.

4.3.3 Decision-Making Through CPS

Acknowledging CPS as a foundational technology for DT is important, especially in decision-making. CPS provides the structural and functional basis through which physical and digital systems can communicate, exchange data, and respond to changing conditions in a coordinated way. In manufacturing research, one important strand has used DT to simulate CPS-based manufacturing systems. For example, Uhlemann, Lehmann and Steinhilper (2017) proposed a Cyber-Physical Production System to improve transparency in production systems and support real-time production control. This illustrates how CPS can strengthen operational visibility and provide a more reliable basis for timely and informed decisions.

Although CPS has not yet been studied as extensively in construction as in manufacturing, some relevant applications are already evident in the built environment. These include energy monitoring systems, monitoring of temporary structures, and mobile crane operations. One notable example is the CPS-based approach proposed by Kan et al. (2018) for planning and monitoring mobile crane deployment and operation on construction project sites. The aim of this approach was to improve the safety of mobile crane operations. The authors proposed a five-layer framework consisting of object, twin, communication, sensing, and actuation layers. In this framework, the object layer represents the physical world, while the actuation layer represents the virtual model. Through CPS, information can be exchanged in both directions, allowing the virtual representation of a crane to be closely coordinated with its physical counterpart on the project site (Kan et al., 2018). This two-way exchange creates a stronger basis for monitoring, coordination, and decision-making in dynamic construction environments.

The opportunity to improve on the CPS phenomenon can be more fully realised through the implementation and use of DT. DT strengthens CPS by enabling more advanced real-time monitoring, simulation, and synchronisation of operational parameters within the system. Operational data such as environmental conditions, physical properties, and system states can be transmitted to the DTS in real time (Wang et al., 2020). This allows the digital system not only to reflect the current condition of the physical system but also to support interpretation and response based on continuously updated information.

DT also integrates historical information with accurate, real-time operational data collected across different stages of a project into a single digital replica. This creates an opportunity for deeper synthesis, interpretation, and evaluation of the status of equipment, assets, or project activities (Tchana,

Ducellier and Remy, 2019a). In the built environment, this is particularly valuable because project conditions often change rapidly, and delayed decisions may lead to inefficiency, disruption, or safety risks. By integrating current and historical information, DT provides a stronger basis for well-informed decisions that can help avoid potential risks and improve operational outcomes.

The use of DT can also improve the decision-making autonomy of the entire system. With better real-time visibility, closer synchronisation, and more integrated analysis, building and construction systems can operate more efficiently and more safely. This has important implications for both construction processes and facility operations. Macchi et al. (2018) noted that more research is still needed to fully explore the implications of implementing DT technology in building construction and operation. Even so, Kan and Anumba (2019) concluded that more cost-effective and efficient building construction and operational processes are possible through DT because it can collect and analyse data at different scales, thereby contributing to a safer environment for building construction and operation.

Decision-making through CPS becomes more powerful when enhanced by DT. While CPS provides the framework for interaction between physical and digital systems, DT extends this capability through richer data integration, deeper analysis, and more intelligent response. For the built environment, this means stronger support for safe, efficient, and timely decisions across both construction and operational settings.

4.3.4 Building Efficiency with Digital Twin

In addition to the safety-related advantages enabled by the real-time tracking functionality of DT, several industries are also exploring its implementation for its strong potential to improve efficiency (Rosen et al., 2015). Efficiency is a critical concern in construction, especially where projects are expected to be completed on time, within budget, and to the required standard of quality (Kaewunruen and Lian, 2019). In the built environment, inefficiency may arise from poor coordination, design clashes, delayed decision-making, inaccurate planning, material waste, low productivity, and weak communication between stakeholders. DT offers an important response to these issues by integrating multi-level data, simulation capabilities, and real-time information into a single connected digital environment. This allows project teams to understand system behaviour more clearly, identify inefficiencies earlier, and improve performance throughout the asset lifecycle. In this way, DT offers new opportunities to improve building efficiency from design through construction to operation.

4.3.4.1 The Use of DT in the Construction Design Process

In the design process, DT can serve as a platform for bringing together a variety of models into one coordinated environment. These models may include conceptual, architectural, and structural models, all of which can be integrated into the

DTS, enabling calculations, simulations, and assessments to be carried out as a unified process (Nie et al., 2019). This is important because design decisions in buildings are rarely isolated. Architectural, structural, environmental, and operational factors often influence one another. DT helps create a common platform for assessing these interrelated components together rather than separately.

The integration enabled by DT improves coordination among system components and supports the creation of simulation scenarios that simultaneously account for multiple design requirements. This means that architectural intent, structural performance, operational implications, and possible future behaviour can all be examined in a more coordinated way. Through such simulations, early design changes can be introduced more effectively, and the likely performance of a proposed building can be predicted with greater confidence (Wang et al., 2020). This improves efficiency because problems can be identified and corrected before they are transferred into the construction stage, where changes are usually more expensive and more disruptive.

4.3.4.2 The Use of DT in the Construction Process

DT also offers several important benefits during the construction process. One of these is the monitoring of construction progress. Through DT, the as-built condition of buildings and facilities can be recreated and updated continuously. This enables real-time comparison of the as-built state with the approved design model. Such comparison helps project teams identify deviations, detect errors, and take corrective action quickly, ensuring the final product remains aligned with the original design intent and specification requirements (Kunz and Fischer, 2020). This supports better quality control and reduces inefficiencies associated with the late discovery of problems.

Another important area is site logistics planning. DT enables the automation and optimisation of resource distribution, equipment deployment, and site usage. As DT can simulate and reflect changing site conditions, it provides a more effective basis for logistics planning. It can help avoid overallocation of resources, reduce unnecessary idle time, and anticipate future on-site resource needs more dynamically (Xu et al., 2018; Kunz and Fischer, 2020). This is especially important on construction sites, where poor logistics often result in delays, congestion, wasted resources, and reduced productivity. More broadly, DT can improve construction efficiency by strengthening coordination among people, processes, equipment, and materials. It provides a clearer view of project status and enables decisions based on current site information rather than assumptions or delayed reports. This contributes to smoother workflow, better sequencing of activities, and more efficient project delivery.

4.3.4.3 The Use of DT in Building Operations

DT is not only useful in design and construction. It can also be used to optimise building operation and to support the long-term management of buildings and facilities. Once a building becomes operational, its efficiency depends on how well its systems are managed, maintained, and adapted to user needs and changing conditions. DT can provide a live digital representation of the building during operation, allowing building managers and operators to monitor performance, assess conditions, and make more informed management decisions (Negri, Fumagalli and Macchi, 2017).

In this sense, DT becomes a valuable tool for improving operational efficiency. It supports the management of building services, the anticipation of performance issues, and the planning of maintenance and operational interventions. It also helps managers prepare for unexpected events by offering a platform for simulation, scenario testing, and forward planning. For this reason, DT can be regarded as an essential tool for maximising a building's potential and for planning effectively for both expected and unexpected operational demands (Kan and Anumba, 2019; Kunz and Fischer, 2020).

DT improves building efficiency by strengthening coordination, supporting simulation, improving monitoring, and enabling better decision-making across the design, construction, and operational phases of built assets. Its value lies in its ability to connect information, anticipate issues, and support more efficient action throughout the building lifecycle.

4.3.5 Digital Twin for Facility Management

The field of building and facility management has long attracted significant interest from many professional disciplines, including architecture, cartography, economics, engineering, law, and management. This is because a building, as an asset, can be understood from many different perspectives. It is not merely a physical structure; it also carries legal, economic, institutional, and managerial value. In this sense, buildings should not be viewed solely as physical entities but also as platforms through which organisational and management objectives can be achieved. This broader understanding is central to facility management, where the focus is not only on the physical condition of a building but also on how effectively it supports the people, services, and activities within.

From an economic perspective, facility management is closely linked to productivity and organisational performance. Noor and Pitt (2009) reported that the intention of facility management includes improving productivity and organisational revenue. This is achieved by continuously creating opportunities to reduce costs through processes that align with organisational goals. Effective facility management, therefore, requires more than routine building maintenance. It involves strategic planning, operational coordination, performance monitoring, and the timely management of assets and services. Araszkiewicz (2017) also noted that effective building and facility management usually involves many professionals who must work together in a coordinated manner to achieve common facility management objectives.

Traditionally, a wide range of professionals, managers, technicians, and artisans from different specialised disciplines are involved in facility management activities. Given the large number of people involved, efficient systems for information exchange and management are necessary to effectively manage buildings and facilities (Scupola, 2012). The incorporation of information technology into facility management has therefore created several opportunities. These include more detailed analysis of daily building operations, quicker identification and estimation of defects, better risk management practices, improved estimation of possible system failures, and stronger protection of life and property (Aziz, Nawawi and Ariff, 2016b).

Over time, various systems have been introduced to support facility management practice. These include the Building Automation System, Computer-Aided Facility Management (CAFM), Computerised Maintenance Management System (CMMS), Energy Management System, and Electronic Document Management System (Yu, Froese and Grobler, 2000). Practical and research evidence have confirmed the usefulness of these systems. However, many of them still have limitations, especially regarding seamless information exchange across systems and stakeholders. This limitation often delays decision-making and may lead to unnecessary actions, delayed interventions, or avoidable inaction. For this reason, although useful, these Information and Communication Technology (ICT)-aided systems are not always sufficient for fully integrated and efficient facility management (Araszkiewicz, 2017).

The development of BIM marked an important turning point for the building sector, particularly in facility management (Aziz, Nawawi and Ariff, 2016a; Aziz, Nawawi and Ariff, 2016b). BIM improved the way building information could be created, stored, visualised, and shared. However, DT takes this progression further. The logic behind DT is built on previous concepts and technologies, including BIM, sensing technologies, and data-driven decision systems. A widely accepted view is that DT can be understood as an advanced form of BIM supported by internet-connected sensors that enable the model to receive real-time updates from the building through continuous data transfer (BSI Group, 2019). For this data to be useful, it must also be analysed using analytics and sometimes artificial intelligence to enable better decisions.

For facility management, DT offers considerable value. It can support operational decision-making, maintenance planning, remodelling, refurbishment, upgrading, and ongoing asset monitoring. A DT platform can also serve as a communication interface among stakeholders at different stages of the facility lifecycle. This is especially important in facility management, where decisions often depend on timely coordination between owners, managers, service providers, maintenance teams, and users. DT can therefore improve communication, reduce unnecessary red tape, minimise rework, and lower energy demand by enabling more connected and informed building management (Alonso et al., 2019). In this regard, DT does not simply digitise facility management. It strengthens it by making the management of buildings more integrated, more intelligent, and more responsive to both operational and organisational needs.

4.3.6 Digital Twin for Building Lifecycle Management

A DT creates virtual intelligent objects from the real world. These virtual intelligent objects make it possible for operating systems and software platforms to track, interpret, and reflect physical components. Through such objects, DT can be used to project the future state of physical assets and simulate how processes may change over time. It is therefore useful when answering 'what if' questions during decision-making (Negri, Fumagalli and Macchi, 2017). DT also allows developers and asset managers to use the data they already possess, including design data, sensor data, and operational records, to combine machine learning, analytics, and other advanced technologies. In doing so, DT creates a bridge between the digital and physical worlds, allowing users to understand how physical assets perform through an interactive and holistic representation of the real environment (Alonso et al., 2019).

In the built environment, DT is frequently used to model building components, systems, and operations. It combines operational technology, which includes operational data and information from building management systems, with information technology in order to provide a seamless flow of data and prompt feedback on a building's operational and informational needs (Kan and Anumba, 2019). This makes DT especially valuable for building lifecycle management because every building passes through a sequence of stages, from conception to eventual renewal, repurposing, or demolition. Every infrastructure asset begins with an idea. It often starts with a developer, owner, or visionary identifying an opportunity either to transform an existing structure or to create something entirely new (Kaewunruen, Sresakoolchai and Zhou, 2020). From that point, the lifecycle progresses through conceptual design, engineering design, construction, commissioning, operation, maintenance, and, eventually, refurbishment, major remodelling, or demolition. In many cases, this end stage then gives rise to a new cycle of development (Yang et al., 2018).

When DT is present across each stage of this lifecycle, it becomes possible to unlock capabilities that traditional digital systems have not fully delivered. Historically, buildings have been digitally modelled during the design stage. However, the digital representation created in one phase has often been ignored or left behind in the next. As a result, data created during design have not always been integrated into commissioning, and commissioning data has not always been kept current during operation (Xu et al., 2018). This creates fragmented and outdated digital records of buildings. When this happens, important information needed for effective decision-making is lost, delayed, or rendered unusable. This is where the full value of DT becomes evident. In an ideal situation, the DT of an asset would be developed even before the physical entity is created (Johnson Controls, 2018).

Long before a building is constructed, the architect and design team produce an initial concept of the structure. Unlike traditional paper-based plans, a digital design can be brought to life immediately and used to simulate the future physical building (Tchana, Ducellier and Remy, 2019a). DT can then update and recompute itself in response to changes in the original drawings, plans, and assumptions. It can assess inherent physical characteristics of the building, such as airflow behaviour and energy use, and it can also evaluate how the building may affect its surrounding environment. For example, it can be used to forecast how a new development may affect traffic flow or surrounding infrastructure. Before the final digital model is used to guide the development of the physical structure, it may undergo hundreds or even thousands of modifications and millions of simulations (Negri, Fumagalli and Macchi, 2017; Kan and Anumba, 2019). This shows the depth of testing and refinement that DT can support before actual construction begins.

For existing buildings, the situation is slightly different. In such cases, the DT often emerges years after the physical asset has already been built and put into service. At first, the DT may only be a rough approximation of the physical asset. However, as more data are collected and additional real-time sensors are installed, the digital and physical twins begin to align more closely in fidelity, detail, and accuracy (Tao and Zhang, 2017). This makes DT especially useful not only for new buildings, but also for existing facilities that need better management, more accurate data, and improved operational insight.

In this regard, DT goes beyond BIM. BIM is a method and a process for creating and managing digital representations of the physical and functional characteristics of places. At the same time, DT is a real-time virtual representation that remains continuously linked to the physical asset (Jupp, 2013). BIM can provide an important information base, but DT extends that capability by drawing from wider building systems, sensor networks, and live operational data. DT can therefore act as a repository for BIM data while also connecting with a far broader ecosystem of smart building and sensing technologies. A real DT is characterised by systems-of-systems integration. As shown in Figure 4.2, the systems integrated into a DT for building lifecycle management may include the following:

- **Embedded sensors**: DT can be created for individual pieces of equipment or for complete systems through 3D modelling software. Embedded sensors then transmit data from the system to administrators, allowing remote monitoring and remote operation. This becomes especially effective where smart equipment is used (Cai et al., 2017).
- **Wireless sensor networks**: These can be installed in buildings to collect data on variables such as temperature, humidity, lighting, and occupancy. Wireless sensor technology makes it easier to expand the system and provides greater flexibility in sensor placement as space needs change over time (Tao et al., 2018).
- **Digitised building lifecycle data and systems**: DT can be created using smart building components, BIM data, and other lifecycle-related information sources. This makes it possible to connect historical and current information in a single environment (Tchana, Ducellier and Remy, 2019a).

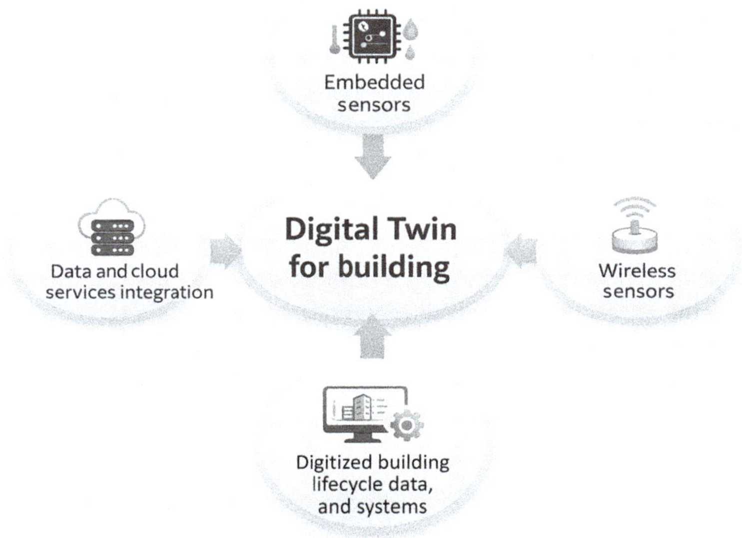

Figure 4.2 DT data sources.

- **Integration with other cloud services and data providers**: External data can be incorporated into a DT through cloud-to-cloud connections, thereby making the system more useful and more intelligent. When buildings lack a full sensor network, cloud-based information from third-party providers can still be used if the structural systems do not connect directly (Alam and El Saddik, 2017).

One of the major challenges facing the global construction industry on its journey towards becoming fully digital is the lack of a consistent data model for building-based applications and systems (Xu et al., 2018). This affects how data are exchanged, interpreted, and reused across lifecycle stages. Even so, DT presents a more advanced response because it not only mimics equipment or devices but also reflects the wider built environment and the immediate surroundings of those assets, thereby enabling more informed judgement and a more timely response to user needs (Nie et al., 2019). For example, the DT of a parking ramp can interact with the DT of a building's security system. Through such interconnection, security cameras may anticipate the arrival of an employee's car, and access may be granted through facial recognition once the driver has been identified in the parking structure (Johnson Controls, 2018). This illustrates how DT can support integrated, responsive, and intelligent building lifecycle management in practice.

4.3.7 From BIM to Digital Twin

In the built environment, DT is often regarded as a major step forward from BIM. There are clear similarities between the objectives of the two concepts, particularly regarding information integration, coordination, and improved decision-making. BIM was developed for the building and construction industry to integrate construction data into a digital environment during project implementation. Its purpose was to support effective collaboration among building professionals and improve the management of 3D information related to buildings and infrastructure (Garber, 2014). In this sense, BIM represented a major shift from fragmented and document-based working methods towards a more coordinated digital approach.

BIM involves the virtual representation of a facility's individual components and how they interact with one another and with the external environment. As the model is prepared for professionals involved in the construction, operation, and eventual deconstruction of a facility, it contains a wide range of data and attributes relating to the physical building. Such information may include construction materials, techniques, costs, dimensions, and other relevant physical and functional characteristics of the asset (Alonso et al., 2019). This makes BIM a valuable information resource for the building lifecycle, especially where collaboration and coordinated information management are important.

Once the project is completed, the information contained in the BIM model is typically handed over to the project owner and, by extension, to facility managers. This large body of facility-specific information can serve as a valuable source for decision-making throughout the lifecycle of the building or infrastructure asset (Boyes, Ellul and Irwin, 2017). The information incorporated in BIM is usually arranged within defined parameters, and changes to the model can be tracked relatively easily. The model can also be produced and managed through architectural information technology tools within common data environments (Alonso et al., 2019). In situations where the required information is not readily available, advanced scanning technologies such as Light Detection and Ranging (LiDAR) can be used to replicate the physical attributes and geometry of a building for the BIM platform. Boyes, Ellul and Irwin (2017) referred to this as the scan-to-BIM process. Through such technologies, more accurate information can be generated for maintenance, operations, and management decisions, thereby encouraging a more data-driven management style (Deutsch, 2015).

The Centre for Advanced Spatial Analysis study further examined the integration of LiDAR and BIM data with real-time data from various IoT sensors and with facility management systems. Through this form of integration, multiple data streams can be combined to create an operational DT for a chosen facility (Dawkins, Hudson-Smith and Dennett, 2018). This is one of the key points at which BIM begins to transition towards DT. BIM provides a structured, information-rich digital foundation, while real-time sensing, analytics, and interconnected systems enable that model to evolve into a live, responsive digital counterpart.

Recent trends in the integration of ICT-powered sensors into built-environment facilities have created greater opportunities to obtain real-time data on building performance, maintenance, and management. When such data are subjected to suitable analysis and simulation, it becomes useful for improving the operational efficiency of the systems and components that make up the built environment. It is important to note that smart and intelligent buildings, and by extension smart and intelligent cities, depend on timely access to reliable data and processed information to function effectively. For this reason, it is reasonable to assert that DT is important for achieving smartness in the built environment in a more efficient and sustainable way (Cousins, 2017).

Applying DT to the built environment can enhance the design process, optimise building operations in relation to the surrounding environment, and improve the prediction of failure and fatigue within building systems and components (Kan and Anumba, 2019). At present, numerous studies continue to explore improved modelling and simulation of built-environment assets, and the outcomes of these efforts are expected to strengthen DT application in the sector (Ito, 2019). In construction engineering, digital building representation has gradually evolved from two-dimensional and 3D modelling to BIM. However, it is important to note that BIM underpins DT in the built environment (Ito, 2019). DT does not replace BIM entirely. Rather, it extends it by introducing real-time data exchange, continuous updating, deeper simulation, and stronger operational intelligence. In this regard, DT can be understood as the progression of BIM from a rich digital model to a living and responsive digital system.

4.3.7.1 Information Sharing Through BIM

It is common practice in the construction industry to exchange and share information among project stakeholders. This is necessary because construction projects involve many participants, including clients, consultants, contractors, subcontractors, suppliers, and facility managers. When information does not move efficiently between these parties, project performance suffers. Issues such as delays, cost overruns, poor coordination, and avoidable errors have often been linked to traditional construction processes and fragmented communication methods (Davidson and Skibniewski, 1995). For this reason, the search for better ways to share and manage information has remained a major concern in the architecture, engineering, and construction (AEC) industry.

With the growth of research and professional discussion on BIM, it has become widely accepted that the concept can have a profound effect on technological advancement and procedural improvement within the industry (Alwisy et al., 2019). BIM provides a digital environment in which building information can be created, managed, and shared in a more structured and coordinated way. Its value lies not only in producing 3D models but also in supporting collaboration among the different parties involved in a project. This explains why BIM has received significant international attention. For example, since 2016, the United Kingdom (UK) government has required all public-sector projects to use fully collaborative 3D BIM technologies and processes, regardless of project size. As a result, BIM has become a subject of major interest in both the UK and the wider global building industry (Kaewunruen and Xu, 2018).

A considerable amount of effort has gone into making collaboration in the AEC industry more efficient, and BIM has contributed strongly to this progress. Olatunji and Sher (2010) noted that engineering teams using BIM found it easier to generate value in the early stages of collaboration. This is because BIM allows information to be developed and shared in a common digital space rather than through isolated drawings and disconnected documents. Over the past few decades, many studies and industry applications have validated the benefits of BIM technology and demonstrated its ability to improve construction processes. Faghihi (2014), for example, demonstrated that BIM data can be used to develop innovative approaches to construction sequencing for installing project components. Since then, numerous studies have shown that digital models can support the continuous, efficient availability of project information throughout the construction process.

Trebbe et al. (2015) documented case studies from a major Dutch railway station rehabilitation project and demonstrated the advantages of using a four-dimensional (4D) CAD system for building projects. Similarly, Lin (2015) proposed a web-based interface management solution for building projects that can help stakeholders improve information sharing and efficiency tracking. These developments show that BIM is not only useful for modelling. It is also valuable as an information-sharing platform that supports collaboration, coordination, sequencing, and communication among project participants.

DT is now adding a new dimension to this discussion by providing a higher level of insight and interaction for stakeholders involved in building projects. One-way DT builds on BIM is by using the digital model as a basis for simulation, analysis, and performance optimisation (Ito, 2019). BIM is an established process that uses 3D modelling to visualise the design, construction, and operation of a building. It is highly useful for information sharing and stakeholder collaboration because it allows different project participants to work from a common platform. However, BIM is largely static and does not fully account for changes that occur over time once the physical asset begins operating. This is where DT becomes important (Khajavi et al., 2019). DT takes BIM a step further by creating a dynamic and interactive representation of the building or asset. It is a virtual replica continuously updated with data from sensors and other information sources, providing real-time performance feedback. Through DT, stakeholders can monitor and optimise different aspects of the building, including energy consumption, maintenance schedules, environmental conditions, and system behaviour (Love and Matthews, 2019). In this way, DT extends information sharing beyond the design and construction stages and into the asset's operational life.

The integration of BIM and DT therefore provides a more comprehensive solution for construction projects. BIM can support the design and construction stages, while DT can strengthen the operation and maintenance stages.

This creates a more seamless transition between project delivery and asset operations, thereby improving communication, collaboration, and information continuity across the building lifecycle (Jung, 2017). DT can also be used to simulate different scenarios and test design options before construction begins, helping stakeholders identify possible issues early and optimise building performance in advance. In addition, it can serve as a platform for training and learning, allowing users to experience various scenarios in a safe, controlled environment (BSI Group, 2019). BIM remains a powerful tool for information sharing in the built environment, while DT builds upon this foundation by introducing real-time updating, live system feedback, and continuous operational intelligence. Together, BIM and DT create stronger opportunities for collaboration, better informed decision-making, and more efficient building delivery and management.

4.3.7.2 BIM Contributions to DT

BIM is a process that involves creating and managing digital representations of the physical and functional characteristics of buildings and other infrastructure. It has become increasingly important in the construction industry and is now widely regarded as a major stepping stone towards the development of DT (BSI Group, 2019). This is because BIM provides a structured, information-rich digital foundation on which more advanced digital systems can be developed. In the built environment, DT does not emerge in isolation. It builds on the data structure, coordination logic, and digital representation already established through BIM. BIM provides a comprehensive 3D model of a building, including its geometry, spatial relationships, and the attributes of building components. In addition to geometric information, BIM models can include other project and asset data, such as performance data, cost data, material specifications, and maintenance schedules. This makes BIM useful not only during design and construction, but throughout the entire lifecycle of a building, from early planning to operation, refurbishment, and eventual demolition (Alonso et al., 2019). In this respect, BIM already contains many of the information structures needed for DT. What DT adds is a stronger link to real-time data, dynamic feedback, simulation, and operational intelligence.

BIM contributes to DT in several important ways. One of the most significant is accurate representation. BIM provides an accurate, detailed representation of the physical building, which is essential for developing a meaningful DT. As the DT depends on a reliable model of the physical asset, the quality and detail of BIM data become very important. The BIM model can be used to simulate and analyse different scenarios, thereby supporting the optimisation of building performance and creating a strong base for the later development of DT (Jung, 2017). Another important contribution is improved collaboration. BIM enables collaboration among the various stakeholders involved in the construction process, including architects, engineers, contractors, and managers. This collaboration supports better communication, stronger coordination, and more consistent information flow across project stages (Kunz and Fischer, 2020). Since DT also depends on coordinated information and stakeholder interaction, BIM contributes by establishing a collaborative environment that can later be extended into a DT framework.

BIM also contributes to risk reduction. It can help identify and resolve potential issues before they occur during construction and operation. This reduces the likelihood of errors, delays, clashes, and cost overruns (Alonso et al., 2019). This early detection of problems is important because DT builds on the use of digital systems to prevent issues before they affect the physical asset. In this sense, BIM supports the preventive logic that DT later advances through real-time updates and predictive capabilities. Another major contribution of BIM is enhanced sustainability. BIM can be used to analyse the environmental performance of a building, including energy consumption and resource use. This helps identify areas where performance can be improved, thereby supporting cost savings and reduced environmental impact (Kaewunruen and Lian, 2019). Since DT is also used to optimise performance during operation, BIM provides an early basis for sustainability analysis that can later be expanded through live monitoring and dynamic optimisation.

BIM further contributes through increased efficiency. It can improve the efficiency of the construction process by reducing waste, improving coordination, and optimising resource use. This can reduce both construction time and cost (Kaewunruen and Lian, 2019). These efficiency gains are important because they show how BIM already supports more intelligent project delivery, while DT extends that intelligence into real-time operation and lifecycle management.

BIM is a critical stepping stone towards DT. It provides the comprehensive digital model and structured information environment needed to create an accurate digital replica of a building or infrastructure asset. The DT can then build on this foundation to support simulation, analysis, optimisation, real-time monitoring, and training. By using BIM as a pathway towards DT, construction professionals can reduce risk, improve collaboration, and strengthen sustainability and efficiency. For this reason, BIM remains one of the most important enabling foundations for DT in the built environment.

4.3.7.3 The BIM - DT Process

The concept of BIM involves the accurate development of a 4D or 5D model for a proposed building or facility. In this model, all major attributes of the proposed construction project are incorporated, while the extent of each attribute is defined by its associated physical boundaries and project requirements (Jones and Bernstein, 2014). BIM creates a digital representation of individual building components and their interaction with the surrounding environment. All relevant physical information, constraints, and characteristics of the building or facility are included in the model. As the model is collectively developed by different built environment professionals through the use of sophisticated architectural ICT tools, changes can be tracked and coordinated more effectively (Jung, 2017).

Recent practice has also seen the introduction of BIM-integrated complementary technologies such as Geographic Information Systems (GIS), CAFM, Construction Project Management Software, and CMMS (Alonso et al., 2019). These technologies have strengthened the digital management of buildings and facilities by expanding the kinds of information that can be captured, organised, and shared. Even so, BIM remains primarily a structured digital modelling and information management system. BIM models the design of buildings and facilities by digitally representing individual building components and their relationships with one another and with the surrounding environment. In this way, all physical information and constraints relating to the building or facility can be incorporated into the model (Alonso et al., 2019). This makes BIM highly valuable during planning, design, and construction. However, one major limitation of BIM is that it does not provide real-time information for immediate operational action. Its value lies mainly in structured information modelling and coordination, but not in continuous live response to changing building conditions.

DT moves beyond this limitation. Unlike BIM, which is largely centred on the building as a designed and documented object, DT is a more intelligent and advanced representation that focuses strongly on the interaction among humans, systems, and buildings through real-time data collection and transmission related to the building's current condition. Over time, the DT of a building evolves in response to physical changes throughout its lifecycle (Mateev, 2020). This makes DT more dynamic, responsive, and operationally relevant than BIM alone.

In the built environment, DT extends across the design, construction, and operational stages of a building. For this reason, Rogers (2019) described DT as 'BIM on steroids', suggesting that its possibilities in the built environment are extensive. Despite the sophistication of BIM, a significant amount of construction and operational information remains beyond its direct scope. DT addresses this gap by introducing dynamic elements, including real-time updates, live system feedback, and automated building functions. These elements include the integration of IoT in buildings and infrastructure, as well as automation within construction and building management systems. Through these features, DT supports the collection, collation, storage, analysis, simulation, and dissemination of critical building information in a more continuous and intelligent manner (Alonso et al., 2019).

To wrap up, the BIM–DT process should be viewed as a progression, not a replacement. BIM provides a structured digital foundation for the asset, while DT enhances it through live data exchange, system intelligence, and real-time interaction. In this way, BIM creates the digital model, and DT activates it into a living, evolving, and continuously informed system. This progression is central to the digital transformation of built assets and the future of smarter building management.

4.4 Benefits of Digital Twin for the Built Assets

DT has gained increasing attention in the built environment for its ability to create a virtual replica of a physical asset or building and connect that replica to real-time and historical data. Through this capability, DT supports real-time monitoring, analysis, simulation, and optimisation of building performance. In practical terms, this means buildings and facilities can be managed more informally, responsively, and efficiently. The technology offers several important benefits, including improved efficiency, reduced cost, enhanced performance, stronger safety management, and increased sustainability (Johnson Controls, 2018). As the built environment becomes more complex and more dependent on data-driven management, the value of DT continues to grow.

One major benefit of DT lies in its application to building design and construction. DT enables architects, engineers, and construction managers to create virtual models of buildings and facilities through which different design options can be visualised, tested, and refined. It also enables simulation of different construction methods, materials, and project scenarios before actual construction begins. This improves the quality of decision-making during the early stages of a project and helps reduce the likelihood of costly design errors, rework, and inefficiencies later in the project lifecycle (Attaran and Celik, 2023). In this regard, DT contributes to better planning, stronger coordination, and more confident project delivery.

Another important benefit is seen in building operation and maintenance. DT provides a virtual representation of the building that can be continuously updated with data from building systems, such as Heating, Ventilation, and Air Conditioning, lighting, and security. This allows operators and facility managers to monitor the condition and performance of building systems in real time. It also supports predictive maintenance by helping managers identify faults, inefficiencies, or likely failures before they become critical. As a result, maintenance can be planned more effectively, downtime reduced, and operational performance improved (Xu et al., 2018; Tchana, Ducellier and Remy, 2019b). For built assets that require long-term maintenance and coordinated operation, this is a major advantage.

DT also makes a significant contribution to energy efficiency and sustainability. By monitoring building performance and analysing energy-related data in real time, building operators can identify areas where energy use can be reduced and system efficiency can be improved. This can result in substantial cost savings and help reduce greenhouse gas emissions. In an era when sustainable building performance is increasingly important, DT offers a platform for more efficient and environmentally responsible building management (Po-Chi and Bonnie, 2011; Alonso et al., 2019). The ability to link operational data to performance analysis also enables sustainability measures to be evaluated more accurately over time.

More broadly, DT provides several benefits for built assets. One of these is improved efficiency. DT allows building operators to optimise systems and operations using real-time feedback and continuous analysis, leading to greater efficiency and reduced operational cost (Kan and Anumba, 2019). It also supports enhanced performance by providing timely data and insights that help operators detect and address issues before they become critical. This contributes to more reliable and better performing buildings (Khajavi et al., 2019). Another benefit is increased sustainability, because DT helps operators monitor and assess energy use and environmental performance more effectively, enabling them to identify areas for improvement and reduce environmental impact (Po-Chi and Bonnie, 2011; Shim et al., 2019). In addition, DT contributes to improved safety. It can be used to simulate emergency situations, test response strategies, and improve the preparedness of building systems and occupants, thereby strengthening safety and security within the built environment (Zhou et al., 2019).

DT provides a wide range of benefits for built assets. Its value lies in its ability to connect physical buildings with intelligent digital systems that support better planning, stronger operational control, improved maintenance, lower cost, increased sustainability, and enhanced safety. As digital transformation in the built environment continues to advance, DT is likely to become an increasingly important tool for achieving smarter, more efficient, and better performing built assets.

4.5 Challenges of Implementing Digital Twin for the Built Assets

Despite the many benefits of DT, several challenges still need to be addressed before it can be adopted more widely across the built environment. The implementation of DT in buildings and infrastructure is not a simple technological upgrade. It requires the coordination of data, systems, people, processes, and long-term management structures. As built assets are complex, long-life systems involving many stakeholders, the transition to DT can be difficult if the necessary technical, organisational, and financial conditions are not in place. Some of the major challenges are discussed below.

One of the most important challenges is data integration. DT depends on integrating multiple data sources, including building systems, sensors, equipment, control systems, maintenance records, and other operational information. In the built environment, these data sources often come from different platforms and may be stored in different formats. Ensuring the accuracy, consistency, and compatibility of this information is crucial for the successful implementation of DT (Attaran and Celik, 2023). If the data are fragmented, incomplete, or poorly coordinated, the DT will not function effectively. This challenge is particularly significant in existing buildings, where legacy systems and inconsistent records often complicate data integration.

Another major challenge is data security. DT involves the collection, transfer, processing, and storage of sensitive information, including building systems data, operational records, and sometimes occupant-related information.

As this data may relate to the building's performance, use, and security, protecting it is essential. Ensuring data privacy and system security is critical to preventing cyber-attacks, unauthorised access, and data breaches (Khajavi et al., 2019). In a built environment increasingly shaped by connected devices and smart systems, weaknesses in digital security can expose both assets and users to significant risk.

Cost is also an important challenge. The implementation of DT often requires significant investment in digital infrastructure, sensing technologies, software platforms, data systems, and specialised expertise. For many organisations, especially smaller firms or building owners with limited resources, this cost can become a major barrier to adoption (Kan and Anumba, 2019). In addition to the initial implementation cost, there are ongoing costs for system updates, maintenance, cybersecurity, and data management. This means that the long-term financial commitment required for DT can be considerable.

Another challenge is DT complexity. Creating and maintaining a DT requires a complex technical architecture and a broad range of skills. As of this complexity, successful DT implementation depends on expertise in areas such as data analysis, modelling, simulation, system integration, and digital asset management (Attaran and Celik, 2023). The built environment does not always have sufficient access to these skills, which can slow adoption or reduce implementation effectiveness. The challenge is therefore not only technological, but also human and organisational.

DT has numerous applications and benefits in the built environment. From building design and construction to operation and maintenance, the technology provides a virtual representation of physical assets, enabling real-time monitoring, analysis, and optimisation. However, for DT to be implemented successfully at scale, the challenges of data integration, data security, cost, and system complexity must be addressed carefully. Although these barriers are significant, they do not outweigh the value of the technology. Rather, they highlight the need for more strategic planning, stronger digital capacity, and better industry readiness in order to unlock the full potential of DT for built assets.

4.6 Summary

This chapter explored the place of DT within the built environment and showed how the technology is gradually becoming an important part of the digital transformation of built assets. It established that buildings and infrastructure are no longer viewed solely as physical structures but increasingly as intelligent, data-rich systems that require more connected, responsive, and lifecycle-oriented forms of management. In this context, DT was presented as a technology that extends beyond static modelling by establishing a live, evolving connection between physical assets and their digital representations.

The chapter explained how DT can be adapted for use in the built environment through stronger data management, real-time monitoring, and decision-making supported by CPS. It also showed that the technology can improve building efficiency across the design, construction, and operational stages of a project. In addition, the discussion highlighted the relevance of DT to facility management and building lifecycle management, where continuous access to information and better coordination among stakeholders are essential.

The chapter also explored the evolution from BIM to DT. It demonstrated that while BIM provides a crucial digital foundation for representing building information, DT enhances this by incorporating real-time data exchange, continuous updates, simulation, and stronger operational intelligence. The discussion further outlined the main benefits of DT for built assets, including improved efficiency, enhanced performance, increased sustainability, and improved safety. It also recognised key challenges in implementation such as data integration, data security, costs, and system complexity. Ultimately, the chapter showed that DT has significant potential to transform how built assets are designed, managed, and operated.

References

Adeniran, A.A. and Shakantu, W. (2023) 'Digitalization of the South African construction industry: A potential for further future research', *Lecture Notes in Civil Engineering*, 245, pp. 17–27. doi:10.1007/978-3-030-97748-1_2

Aghimien, D., Aigbavboa, C. and Oke, A. (2018) 'Digitalisation for Effective Construction Project Delivery in South Africa', in *Contemporary Construction Conference: Dynamic and Innovative Built Environment*, pp. 3–10.

Aghimien, D. et al. (2022) 'PLS-SEM assessment of the impediments of robotics and automation deployment for effective construction health and safety', *Journal of Facilities Management*. doi:10.1108/JFM-04-2022-0037

Akinshipe, O., Aliu, J. and Aigbavboa, C. (2022) 'The futuristic building industry: Updating skills-set required for digital transformation', *Social and Occupational Ergonomics*, 65. doi:10.54941/AHFE1002657

Alaloul, W.S. et al. (2020) 'Industrial Revolution 4.0 in the construction industry: Challenges and opportunities for stakeholders', *Ain Shams Engineering Journal*, 11(1), pp. 225–230. doi:10.1016/J.ASEJ.2019.08.010

Alam, K.M. and El Saddik, A. (2017) 'C2PS: A digital twin architecture reference model for the cloud-based cyber-physical systems', *IEEE Access*, 5, pp. 2050–2062. doi:10.1109/ACCESS.2017.2657006

Alonso, R. et al. (2019) 'SPHERE: BIM digital twin platform', *Proceedings*, 20(9), pp. 1–5. doi:10.3390/proceedings2019020009

Alwisy, A. et al. (2019) 'A BIM-based automation of design and drafting for manufacturing of wood panels for modular residential buildings', *International Journal of Construction Management*, 19(3), pp. 187–205. doi:10.1080/15623599.2017.1411458

Araszkiewicz, K. (2017) 'Digital Technologies in Facility Management - The state of Practice and Research Challenges', *Procedia Engineering*, 196, pp. 1034–1042. doi:10.1016/j.proeng.2017.08.059

Attaran, M. and Celik, B.G. (2023) 'Digital Twin: Benefits, use cases, challenges, and opportunities', *Decision Analytics Journal*, 6, p. 100165. doi:10.1016/J.DAJOUR.2023.100165

Aziz, N.D., Nawawi, A.H. and Ariff, N.R.M. (2016a) 'Building information modelling (BIM) in facilities management: opportunities to be considered by facility managers', *Procedia - Social and Behavioral Sciences*, 234, pp. 353–362. doi:10.1016/j.sbspro.2016.10.252

Aziz, N.D., Nawawi, A.H. and Ariff, N.R.M. (2016b) 'ICT evolution in facilities management (FM): Building information modelling (BIM) as the latest technology', *Procedia - Social and Behavioral Sciences*, 234, pp. 363–371. doi:10.1016/j.sbspro.2016.10.253

Boyes, G.A., Ellul, C. and Irwin, D. (2017) 'Exploring BIM for operational integrated asset management - A preliminary study utilising real-world infrastructure data', in *ISPRS Annals of the Photogrammetry, Remote Sensing and Spatial Information Sciences*, IV-4/W5, pp. 49–56. doi:10.5194/ISPRS-ANNALS-IV-4-W5-49-2017

BSI Group (2019) 'Digital twins for the built environment'. *BIMToday*. Available at: https://www.pbctoday.co.uk/news/bim-news/digital-twin-4-0/64519 (Accessed: 25 March 2026).

Cai, Y. et al. (2017) 'Sensor data and information fusion to construct Digital-Twins virtual machine tools for cyber-physical manufacturing', *Procedia Manufacturing*, 10, pp. 1031–1042. doi:10.1016/j.promfg.2017.07.094

Cousins, S. (2017) '3D mapping Helsinki: How mega digital models can help city planners', *Construction Research and Innovation*, 8(4), pp. 102–106. doi:10.1080/20450249.2017.1396747

Dall'Omo, S. (2017) 'Driving African development through smarter technology'. African Digitalisation Maturity Report.

Davidson, I.N. and Skibniewski, M.J. (1995) 'Simulation of automated data collection in buildings', *Journal of Computing in Civil Engineering*, 9(1), pp. 9–20. doi:10.1061/(ASCE)0887-3801(1995)9:1(9)

Dawkins, O., Hudson-Smith, A. and Dennett, A. (2018) 'Living with a Digital Twin: Operational management and engagement using IoT and mixed realities at UCL's here east campus on the Queen Elizabeth Olympic Park', *BIM-DigiCons*, p. 6. Available at: https://pdfs.semanticscholar.org/da61/a1bf345892d7504bde14624788ee9dd07b47.pdf (Accessed: 7 January 2020).

Deutsch, R. (2015) *Data-driven design and construction: 25 strategies for capturing, analyzing and applying building data*. Hoboken, NJ: John Wiley & Sons.

Dithebe, K., Aigbavboa, C., Oke, A. and Muyambu, M.A. (2018) 'Factors influencing the performance of the South African construction industry: A case of Limpopo province', in *Proceedings of the International Conference on Industrial Engineering and Operations Management*, pp. 1185–1192.

Faghihi, V. (2014) *Automated and optimized project scheduling using BIM*. College Station, TX: Texas A&M University. Available at: http://hdl.handle.net/1969.1/152576 (Accessed: 8 July 2023).

Garber, R. (2014) 'BIM design: Realising the creative potential of building information modelling', *BIM Design: Realising the Creative Potential of Building Information Modelling*, pp. 1–248. doi:10.1002/9781118914694

Glaessgen, E.H. and Stargel, D.S. (2012) 'The digital twin paradigm for future NASA and U.S. Air force vehicles', in *Proceedings of the 53rd AIAA/ASME/ASCE/AHS/ASC Structures, Structural Dynamics and Materials Conference*. Reston, VA: American Institute of Aeronautics and Astronautics, pp. 1–14. doi:10.2514/6.2012-1818

Grieves, M. and Vickers, J. (2016) 'Digital twin: Mitigating unpredictable, undesirable emergent behavior in complex systems', in Kahlen, F.-J., Flumerfelt, S. and Alves, A. (eds.) *Transdisciplinary perspectives on complex systems: New findings and approaches*. Cham: Springer, pp. 85–113. doi:10.1007/978-3-319-38756-7_4

Hasan, S. (2015) 'Digitalization of the construction industry: Building information modeling'. *Building Radar*. Available at: https://buildingradar.com/construction-blog/building-information-modeling (Accessed: 8 July 2023).

Hosamo, H.H., Nielsen, H.K., Kraniotis, D., Svennevig, P.R. and Svidt, K. (2023) 'Digital Twin framework for automated fault source detection and prediction for comfort performance evaluation of existing non-residential Norwegian buildings', *Energy and Buildings*, 281, p. 112732. doi:10.1016/j.enbuild.2022.112732

Ito, K. (2019) 'Change beyond building information modeling to a "Digital Twin" in architecture', *Japan Architectural Review*, 2(4), pp. 391–391. doi:10.1002/2475-8876.12123

Jazzar, M. E., Urban, H., Schranz, C. and Nassereddine, H. (2020) 'Construction 4.0: A roadmap to shaping the future of construction', in *Proceedings of the 37th International Symposium on Automation and Robotics in Construction (ISARC)*, pp. 1314–1321. doi:10.22260/ISARC2020/0180

Johnson Controls (2018) *Applying Digital Twins to the built environment*. Available at: https://www.johnsoncontrols.com/-/media/jci/insights/2019/bts/jci-661_dv_digital_twin_white_paper_020819_6p_f3.pdf (Accessed: 8 January 2020).

Jones, S.A. and Bernstein, H.M. (2014) 'The business value of BIM for owners', in *SmartMarket Report*. Bedford, MA: McGraw Construction. https://damassets.autodesk.net/content/dam/autodesk/www/solutions/building-information-modeling/bim-value/Business_Value_of_BIM_for_Owners_SMR_(2014).pdf

Jung, D. (2017) 'Why BIM and Digital Twin Technology shouldn't be confused'. *Sablono*. Available at: https://www.sablono.com/en/blog/bim-and-digital-twin-technology (Accessed: 8 January 2020).

Jupp, J.R. (2013) 'Incomplete BIM implementation: Exploring challenges and the role of product lifecycle management functions', in A. Bernard, L. Rivest and D. Dutta (eds.), *Product lifecycle management for society. PLM 2013. IFIP Advances in Information and Communication Technology*, vol. 409. Berlin, Heidelberg: Springer. doi:10.1007/978-3-642-41501-2_62

Kaewunruen, S. and Lian, Q. (2019) 'Digital twin aided sustainability-based lifecycle management for railway turnout systems', *Journal of Cleaner Production*, 228, pp. 1537–1551. doi:10.1016/j.jclepro.2019.04.156

Kaewunruen, S., Sresakoolchai, J. and Zhou, Z. (2020) 'Sustainability-based lifecycle management for bridge infrastructure using 6D BIM', *Sustainability*, 12(6), p. 2436. doi:10.3390/SU12062436

Kaewunruen, S. and Xu, N. (2018) 'Digital Twin for sustainability evaluation of railway station buildings', *Frontiers in Built Environment*, 4, p. 77. doi:10.3389/fbuil.2018.00077

Kan, C. and Anumba, C.J. (2019) 'Digital Twins as the next phase of cyber-physical systems in construction', in *Computing in Civil Engineering: Smart Cities, Sustainability, and Resilience*. Reston, VA: American Society of Civil Engineers, pp. 256–264. doi:10.1061/9780784482438.033

Kan, C., Fang, Y., Anumba, C.J. and Messner, J.I. (2018) 'A cyber–physical system for planning and monitoring mobile cranes on construction sites', *Proceedings of the Institution of Civil Engineers - Management, Procurement and Law*, 171(4), pp. 161–174. doi:10.1680/jmapl.17.00042

Khajavi, S.H. et al. (2019) 'Digital Twin: Vision, benefits, boundaries, and creation for buildings', *IEEE Access*, 7, pp. 147406–147419. doi:10.1109/ACCESS.2019.2946515

Knapp, G.L. et al. (2017) 'Building blocks for a digital twin of additive manufacturing', *Acta Materialia*, 135, pp. 390–399. doi:10.1016/j.actamat.2017.06.039

Kunz, J. and Fischer, M. (2020) 'Virtual design and construction', *Construction Management and Economics*, 38(4), pp. 355–363. doi:10.1080/01446193.2020.1714068

Li, B.H., Zhang, L., Li, T., Lin, T.Y. and Cui, J. (2017) 'Simulation-based cyber-physical systems and internet-of-things', in S. Mittal, U. Durak and T. Ören (eds.), *Guide to simulation-based disciplines. Simulation foundations, methods and applications*. Cham, Switzerland: Springer. doi:10.1007/978-3-319-61264-5_5

Lin, Y.-C. (2015) 'Use of BIM approach to enhance construction interface management: A case study', *Technological and Economic Development of Economy*, 21(2), pp. 201–223. doi:10.3846/13923730.2013.802730

Love, P.E.D. and Matthews, J. (2019) 'The "how" of benefits management for digital technology: From engineering to asset management', *Automation in Construction*, 107, p. 102930. doi:10.1016/j.autcon.2019.102930

Macchi, M. et al. (2018) 'Exploring the role of Digital Twin for asset lifecycle management', *IFAC-PapersOnLine*, 51(11), pp. 790–795. doi:10.1016/j.ifacol.2018.08.415

Madubuike, O.C. and Anumba, C.J. (2022) 'Digital Twin–based health care facilities management', *Journal of Computing in Civil Engineering*, 37(2), p. 04022057. doi:10.1061/JCCEE5.CPENG-4842

Mateev, M. (2020) 'Industry 4.0 and the digital twin for building industry', *International Scientific Journal "Industry 4.0"*, 5(1), pp. 29–32. Available at: https://stumejournals.com/journals/i4/2020/1/29.full.pdf (Accessed: 7 January 2020).

Negri, E., Fumagalli, L. and Macchi, M. (2017) 'A review of the roles of Digital Twin in CPS-based production systems', *Procedia Manufacturing*, 11, pp. 939–948. doi:10.1016/J.PROMFG.2017.07.198

Nie, J. et al. (2019) 'Digital Twin-based smart building management and control framework', in Saika, O. and Cohen, E. (eds.) *Proceedings of the International Conference on Artificial Intelligence and Computing Science (ICAICS)*, Hangzhou, pp. 10–14. doi:10.12783/dtcse/icaic2019/29395

Noor, M.N.M. and Pitt, M. (2009) 'A critical review on innovation in facilities management service delivery', *Facilities*, 27(5–6), pp. 211–228. doi:10.1108/02632770910944943

Oke, A. et al. (2018) 'Challenges of digital collaboration in the South African construction industry', in *Proceedings of the International Conference on Industrial Engineering and Operations Management*, Bandung, Indonesia. Available at: http://ieomsociety.org/ieom2018/papers/608.pdf

Olatunji, O.A. and Sher, W. (2010) 'Legal implications of BIM: model ownership and other matters arising', in *Proceedings of the 18th CIB World Congress*, Salford, UK.

Omer, M. et al. (2019) 'Use of gaming technology to bring bridge inspection to the office', *Structure and Infrastructure Engineering*, 15(10), pp. 1292–1307. doi:10.1080/15732479.2019.1615962

Opoku, D.G.J. et al. (2021) 'Digital twin application in the construction industry: A literature review', *Journal of Building Engineering*, 40, p. 102726. doi:10.1016/J.JOBE.2021.102726

Osunsanmi, T.O., Oke, A.E. and Aigbavboa, C.O. (2018) 'Fusing RFID with mobile technology for enhanced safety of construction project team members', in *Proceedings of the International Conference on Industrial Engineering and Operations Management*, Pretoria, RSA, pp. 312–321. Available at: https://ieomsociety.org/southafrica2018/papers/102.pdf (Accessed: 26 March 2022).

Po-Chi, L. and Bonnie, L.Y.M. (2011) 'An intelligent building index awarded building', *Intelligent Buildings International*, 3(1), pp. 55–62. doi:10.3763/inbi.2010.0005

Re Cecconi, F. et al. (2021) *Soft Digital Twin for IEQ enabling the COVID risk mitigation in educational spaces*. Available at: https://re.public.polimi.it/handle/11311/1187733 (Accessed: 30 April 2023).

RICS (2017) *Associate assessment: Quantity surveying and construction*. UK: Royal Institution of Chartered Surveyors. Available at: https://www.rics.org/content/dam/ricsglobal/documents/join-rics/qs_and_construction_pathway_guide_associate_rics.pdf (Accessed: 25 March 2026).

Rogers, D. (2019) 'Twin tracks: The drive to create a smart digital model of the UK', *Construction Research and Innovation*, 10(2), pp. 49–52. doi:10.1080/20450249.2019.1621591

Rosen, R. et al. (2015) 'About the importance of autonomy and digital twins for the future of manufacturing', *IFAC-PapersOnLine*, 48(3), pp. 567–572. doi:10.1016/j.ifacol.2015.06.141

Schleich, B. et al. (2017) 'Shaping the digital twin for design and production engineering', *CIRP Annals - Manufacturing Technology*, 66(1), pp. 141–144. doi:10.1016/j.cirp.2017.04.040

Scupola, A. (2012) 'ICT adoption in facilities management supply chain: The case of Denmark', *Journal of Global Information Technology Management*, 15(1), pp. 53–78. doi:10.1080/1097198X.2012.10845612

Shahzad, M. et al. (2022) 'Digital Twins in built environments: An investigation of the characteristics, applications, and challenges', *Buildings*, 12(2), p. 120. doi:10.3390/BUILDINGS12020120

Shim, C.S. et al. (2019) 'Development of a bridge maintenance system for prestressed concrete bridges using 3D digital twin model', *Structure and Infrastructure Engineering*, 15(10), pp. 1319–1332. doi:10.1080/15732479.2019.1620789

Söderberg, R. et al. (2017) 'Toward a Digital Twin for real-time geometry assurance in individualized production', *CIRP Annals - Manufacturing Technology*, 66(1), pp. 137–140. doi:10.1016/j.cirp.2017.04.038

Tanga, O.T. et al. (2021) 'Construction digitalisation tools in South African construction industry: An added advantage', *IOP Conference Series: Materials Science and Engineering*, 1107(1), p. 012230. doi:10.1088/1757-899X/1107/1/012230

Tao, F. and Zhang, M. (2017) 'Digital Twin shop-floor: A new shop-floor paradigm towards smart manufacturing', *IEEE Access*, 5, pp. 20418–20427. doi:10.1109/ACCESS.2017.2756069

Tao, F., Zhang, M., Liu, Y. and Nee, A.Y.C. (2018) 'Digital twin driven prognostics and health management for complex equipment', *CIRP Annals - Manufacturing Technology*, 67(1), pp. 169–172. doi:10.1016/j.cirp.2018.04.055

Tchana, Y., Ducellier, G. and Remy, S. (2019a) 'Designing a unique Digital Twin for linear infrastructure lifecycle management', *Procedia CIRP*, 84, pp. 545–549. doi:10.1016/j.procir.2019.04.176

Tchana, Y., Ducellier, G. and Remy, S. (2019b) 'Designing a unique Digital Twin for linear infrastructures lifecycle management', in *29th CIRP Design*. Elsevier B.V., pp. 545–549. doi:10.1016/j.procir.2019.04.176

Trebbe, M., Hartmann, T. and Dorée, A. (2015) '4D CAD models to support the coordination of construction activities between contractors', *Automation in Construction*, 49, pp. 83–91. doi:10.1016/j.autcon.2014.10.002

Uhlemann, T.H.J., Lehmann, C. and Steinhilper, R. (2017) 'The Digital Twin: Realizing the cyber-physical production system for Industry 4.0', in *Proceedings of the 24th CIRP Conference on Life Cycle Engineering*, pp. 335–340. doi:10.1016/j.procir.2016.11.152

Wang, Y. et al. (2020) 'Big data driven Hierarchical Digital Twin Predictive Remanufacturing paradigm: Architecture, control mechanism, application scenario and benefits', *Journal of Cleaner Production*, 248, p. 119299. doi:10.1016/j.jclepro.2019.119299

Xu, Y. et al. (2018) 'Digital Twin concept in managing industrial capital construction projects life cycle', in *Proceedings of the 11th International Conference Management of Large-Scale System Development (MLSD)*. IEEE. doi:10.1109/MLSD.2018.8551867

Yang, X. et al. (2018) 'Building-information-modeling enabled life cycle assessment, a case study on carbon footprint accounting for a residential building in China', *Journal of Cleaner Production*, 183, pp. 729–743. doi:10.1016/J.JCLEPRO.2018.02.070

Yu, K., Froese, T. and Grobler, F. (2000) 'Development framework for data models for computer-integrated facilities management', *Automation in Construction*, 9(2), pp. 145–167. doi:10.1016/S0926-5805(99)00002-3

Zhou, C. et al. (2019) 'Cyber-physical-system-based safety monitoring for blind hoisting with the internet of things: A case study', *Automation in Construction*, 97, pp. 138–150. doi:10.1016/j.autcon.2018.10.017

5 Managing Indoor Environmental Quality with Digital Twin

5.1 Introduction

The quality of the indoor environment has become one of the most important concerns in the design, operation, and management of buildings. This is because indoor spaces directly affect the health, comfort, productivity, and general well-being of those who occupy them. As individuals spend most of their time indoors, the conditions within buildings have a strong and lasting influence on daily life, work performance, and quality of experience (Levin, 2006; World Health Organisation, 2010; Cai et al., 2023). For this reason, indoor environmental quality (IEQ) has increasingly gained attention in both academic and professional discussions surrounding healthy and sustainable buildings.

IEQ generally refers to the quality of indoor conditions as experienced by building users. It is commonly understood in terms of four major components: indoor air quality, thermal comfort, lighting, and acoustic quality. These factors interact in complex ways, shaping how occupants perceive and respond to the indoor environment (Frontczak and Wargocki, 2011; Dykes and Baird, 2013). Poor IEQ has been associated with a wide range of negative outcomes, including respiratory illness, discomfort, stress, fatigue, reduced productivity, and broader health concerns (Attia, 2018). At the same time, improving IEQ is not always straightforward. Building managers often face the challenge of balancing comfort, health, energy use, maintenance demands, and operational costs within a single integrated system (Roulet, 2001; Geng, Lin and Zhu, 2020).

The growing complexity of buildings has made conventional approaches to IEQ management increasingly limited. Periodic inspection, manual control, and isolated building systems do not always provide the responsiveness needed to manage indoor conditions effectively. Contemporary buildings require more intelligent and connected systems that can continuously monitor environmental conditions, interpret data meaningfully, and support timely intervention. In this context, the Digital Twin (DT) has emerged as a promising approach. As a virtual representation of a physical asset or system, DT can support real-time monitoring, analysis, prediction, and optimisation of building conditions (van Midden, 2022; Cai et al., 2023). This makes it especially relevant to IEQ

DOI: 10.1201/9781003743330-5

management, where continuous environmental change requires equally continuous digital awareness and control.

A Digital Twin System (DTS) for IEQ management offers a proactive, data-driven approach to understanding and improving indoor conditions. It can integrate information from sensors, building systems, and operational data to provide a more complete view of indoor environmental performance. This supports real-time monitoring, predictive maintenance, system optimisation, benchmarking, and more informed decision-making regarding the indoor environment (Re Cecconi et al., 2021; Cai et al., 2023). In doing so, DT creates an opportunity to move beyond reactive facility management towards a more adaptive and occupant-centred model of building operation.

This chapter explores the concept of a DTS for IEQ management. It begins with an overview of IEQ and its key components, before discussing human perception, improvement strategies, the balance between IEQ and energy use, and the health effects of poor IEQ. The chapter then considers how DT can enhance IEQ management, outlines the benefits of such a system, and discusses the theoretical foundations that support its application. In this way, the chapter provides a clear basis for understanding how DT can contribute to healthier, more comfortable, and more intelligent indoor environments.

5.2 Overview of Indoor Environmental Quality

One of the major purposes of a building is to provide an indoor environment that is healthy, safe, and comfortable for its occupants (Sakhare and Ralegaonkar, 2014; Al-Sulaihi et al., 2015). However, achieving satisfactory IEQ has remained difficult in many contexts. Rapid urbanisation, increasing housing demand, overcrowding, and growing pressure on energy, water, green spaces, and infrastructure have all made the delivery of quality indoor environments more challenging. Even so, the construction industry continues to pursue better ways of improving comfort and IEQ. Contemporary buildings are increasingly designed with sustainability in mind, which has brought greater attention to the indoor environment. This focus is essential because people spend most of their time indoors, making indoor conditions highly important to health, comfort, and daily functioning (Levin, 2006).

IEQ refers to the quality of the indoor environment within buildings. It includes indoor air quality, thermal comfort, lighting, and acoustic quality. IEQ is important because it affects the health, comfort, satisfaction, and productivity of building occupants. According to the World Health Organisation, poor IEQ can contribute to health problems such as respiratory diseases, allergies, eye irritation, and headaches (World Health Organisation, 2010). The importance of IEQ is increasingly being recognised by building owners, facility managers, and public health authorities. As more people live and work in enclosed environments, especially in urban areas, the quality of the indoor environment directly affects well-being and quality of life. The (World Health

Organisation, 2022) also estimated that around 3.8 million premature deaths each year are attributable to indoor air pollution.

Building occupants are exposed to a wide range of indoor pollutants and contaminants. These include volatile organic compounds (VOCs), carbon monoxide (CO), nitrogen oxides, and particulate matter. These pollutants can originate from building materials, furnishings, cleaning products, combustion appliances, and other indoor sources. Poor ventilation and inadequate building maintenance can further worsen these conditions (Environmental Protection Agency, 2022b). For this reason, IEQ remains a critical aspect of both building design and building operation. It is not only a technical issue but also a human-centred concern because it affects the health, comfort, and well-being of occupants. The following sections discuss the major components of IEQ, the health effects of poor IEQ, methods of measuring IEQ, and strategies for improving IEQ.

Several factors influence the quality of the indoor environment in buildings. The most widely recognised components of IEQ are air quality, thermal comfort, lighting, and acoustic quality.

5.2.1 Air Quality

Air quality refers to the composition of indoor air, including the presence of pollutants and contaminants. Indoor air quality may be affected by many factors, including outdoor pollution, building materials, cleaning agents, and combustion appliances. Poor indoor air quality can lead to respiratory illness, allergies, headaches, and other health complaints (Hulin et al., 2012). Several studies have investigated the effect of indoor air pollutants on health. One study found that exposure to pollutants such as formaldehyde and benzene was associated with an increased risk of asthma (Richardson et al., 2005). Another study found that exposure to pollutants such as CO and nitrogen oxides was associated with increased respiratory symptoms and infections (Hulin et al., 2012).

5.2.2 Thermal Comfort

Thermal comfort refers to the level of satisfaction with the thermal environment, including temperature, humidity, and air movement. The American Society of Heating, Refrigerating and Air-Conditioning Engineers (ASHRAE) defines thermal comfort as the condition of mind that expresses satisfaction with the thermal environment (ASHRAE, 2020). The thermal environment can strongly affect well-being and productivity. It has been reported that employees who were satisfied with their thermal environment showed higher levels of productivity (Al Horr et al., 2016b). Thermal discomfort has also been linked to symptoms such as headaches, fatigue, and eye irritation (Lan et al., 2011).

5.2.3 Lighting

Lighting refers to the level and quality of light in the indoor environment. Good lighting can improve visual comfort, task performance, and productivity, while poor lighting can contribute to eye strain, fatigue, and discomfort (Küller et al., 2007). Several studies have explored the relationship between lighting and health. Research has shown that improved lighting conditions can reduce anxiety and stress levels, and can also improve the academic performance of students (Khorshid et al., 2021).

5.2.4 Acoustic Quality

Acoustic quality refers to the level and quality of sound in the indoor environment. Good acoustic quality can improve speech intelligibility and reduce noise-related stress, while poor acoustic quality can lead to annoyance, distraction, and discomfort (Stansfeld and Matheson, 2003). Studies in the UK have shown that high noise levels are associated with increased sleep disturbance. Noise has also been linked to higher levels of self-reported stress and fatigue in open-plan office environments (Khajenasiri et al., 2016).

5.3 Human Perception of Indoor Environmental Quality

Beyond the core components of IEQ, other variables also influence how people perceive comfort, serenity, and satisfaction within indoor spaces. Human perception of IEQ is not shaped only by air quality, temperature, lighting, and sound. It is also influenced by the wider experience of being in a space. Fantozzi and Rocca (2020) noted that workers' productivity and comfort in office environments are affected by eight exogenous factors. These include thermal comfort, visual and daylighting quality, noise and acoustics, indoor air quality and ventilation, space layout, biophilia and views, look and feel, location, and amenities. Together, these factors help explain why human perception of IEQ is broader than that of traditional technical indicators alone. These factors are illustrated in Figure 5.1 below.

In recent years, researchers have examined users' emotional and psychological responses to indoor spaces in greater depth. This has been done in order to make indoor spaces more appealing, reduce stress, and improve user efficiency, satisfaction, and productivity. For example, creative and nature-based design solutions such as living walls and potted plants have increasingly been introduced into workplaces to make them more pleasant and supportive environments (Danza et al., 2020). Experimental evidence has shown that biophilic indoor spaces, whether enhanced by real vegetation or simulated natural elements, can support human well-being. These environments can produce meaningful improvements in both physiological and psychological health. Biophilic workspaces have been associated with improved attentiveness, better stress recovery among staff members, and improved psychological and cognitive performance (Yin et al., 2018; van den Berg et al., 2016).

It is important to note that recent years have also seen growing interest in human multisensory interaction within psychology and neuroscience, even though the full effects of IEQ factors have not yet been completely understood (Yang and Moon, 2019). One of the most interesting issues in this area is how

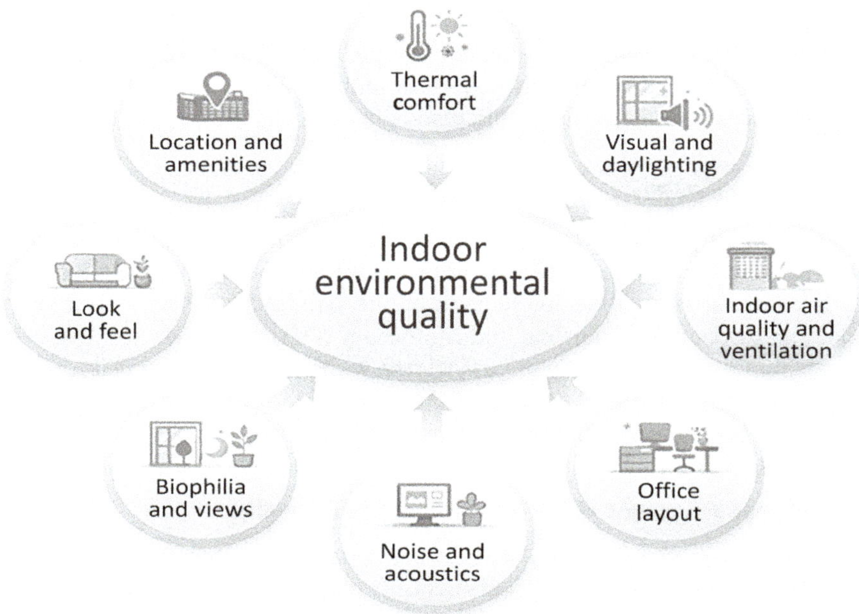

Figure 5.1 Comprehensive human perception of IEQ.

to estimate the extent to which each IEQ indicator affects overall IEQ. The scientific community has not yet reached full consensus on this matter. Findings from different studies, particularly those focusing on air-conditioned office buildings, vary widely. Humphreys (2007), for example, used participant satisfaction data from 26 office buildings across five European countries to rank the factors that affect IEQ. After averaging the responses, thermal comfort and air quality emerged as the most important factors for perceived comfort. This supports the wider view that thermal comfort and indoor air quality strongly affect occupants' satisfaction. However, differences remain when results from individual studies are compared.

A large body of IEQ research has been carried out in different regions of the world, including Australia (Residovic, 2017), Africa (Nimlyat, 2018), China (Wong et al., 2018), Italy (Marino et al., 2012), and the UK (Ncube and Riffat, 2012). Many of these studies support the view that thermal comfort and indoor air quality have the greatest influence on the assessment of environmental quality in indoor spaces. At the same time, some findings differ. Studies conducted in office buildings and analysed by Heinzerling et al. (2013) found that air quality and visual comfort were the most important factors, whereas thermal comfort was the least important factor for IEQ. This contrasted with many earlier studies, which consistently identified thermal comfort as one of the strongest influences. These differences show that human perception of IEQ is not always fixed and may vary depending on context, building type, user expectations, and environmental conditions.

One of the major determinants of IEQ is the building's location and the surrounding environmental characteristics. IEQ is also influenced by building-related factors such as building condition, maintenance quality, and cleaning protocols (US Environmental Protection Agency, 2016). In addition, contaminants generated inside the building can significantly affect IEQ. Indoor spaces may be polluted by mould, allergens, particles, bacteria, formaldehyde, and VOCs (Zuo and Zhao, 2014). The IEQ characteristics of a building also depend on its type, age, and the materials used in construction. In some older buildings, for example, asbestos may still be present, while poor insulation and leak-prone construction may further worsen indoor environmental conditions (Turunen et al., 2014).

Studies by Daisey, Angell, and Apte (2003) noted that some health-related symptoms can be directly traced to inadequate IEQ in buildings. Poor IEQ can compromise the health of building users and occupants. Earlier reports have shown that excessive concentrations of carbon dioxide in enclosed spaces can lead to unpleasant odours, fatigue, dizziness, headaches, and restlessness. Mendell et al. (2013) also reported that adequate ventilation is directly linked to a lower incidence of illness among users of public buildings, such as schools. Studies by Wålinder et al. (2007) found that users often perceive indoor air quality to be worse when indoor temperature rises. This confirms a strong relationship between indoor temperature, perceived stuffiness, and physiological response. In the same way, poor indoor air quality is often reported in public indoor spaces where per-user ventilation is low. In other words, ventilation is directly related to perceived indoor air quality. When ventilation is poor, users are more likely to report poor daily indoor air quality (Haverinen-Shaughnessy and Shaughnessy, 2015).

Human perception of IEQ is shaped by a combination of technical, environmental, sensory, emotional, and contextual factors. This makes IEQ a deeply human-centred issue. Understanding how people perceive indoor environments is therefore essential for designing, managing, and improving buildings in ways that truly support health, comfort, and well-being.

5.4 Strategies for Improving Indoor Environmental Quality

Maintaining good IEQ is essential for promoting occupant health, well-being, comfort, and productivity. Poor IEQ can lead to a wide range of health problems, including respiratory difficulties, headaches, allergies, fatigue, and general discomfort (Tietjen et al., 2012). For this reason, IEQ should be treated as a major priority in the design, operation, and maintenance of buildings. Improving IEQ does not depend on a single action. Rather, it requires a combination of strategies that address air quality, ventilation, thermal conditions, lighting, cleanliness, and acoustic performance in an integrated way. This section discusses some of the most effective strategies for improving IEQ.

1. **Use of natural ventilation**: Natural ventilation refers to the supply and removal of air from an indoor space without relying on mechanical systems. It can improve indoor air quality by removing indoor pollutants and introducing fresh outdoor air. Natural ventilation may be achieved through operable windows, skylights, vents, and other passive ventilation systems (Holm and Engelbrecht, 2005). Studies have shown that natural ventilation can significantly reduce indoor air pollution levels and improve indoor environmental conditions (Goyal and Khare, 2011; Haverinen-Shaughnessy and Shaughnessy, 2015). Ventilation remains one of the most important factors in maintaining acceptable indoor air quality. Proper ventilation helps to remove harmful pollutants from indoor spaces while supplying fresh outdoor air. ASHRAE recommends a minimum of 15 cubic feet per minute of outdoor air per person in commercial buildings (ASHRAE, 2020). In addition to achieving adequate ventilation rates, it is also important to ensure that the ventilation system is properly designed, operated, and maintained so that it does not become a source of indoor air quality problems.
2. **Proper air purification and filtration**: Air filtration is the process of removing contaminants from indoor air. Proper filtration can significantly improve indoor air quality by removing allergens, pollutants, particulate matter, and other unwanted contaminants. The use of high-efficiency air filters within Heating, Ventilation, and Air-Conditioning (HVAC) systems can improve indoor air quality by reducing pollutant levels in occupied spaces (Agbo et al., 2021). Studies have also shown that appropriate air filtration can reduce the risk of respiratory illness and support better health outcomes (Environmental Protection Agency, 2022a). In this sense, air purification and filtration are important strategies for maintaining healthier indoor spaces, especially in areas with greater exposure to pollution or among vulnerable occupants.
3. **Maintenance of HVAC systems**: Proper maintenance of HVAC systems is essential for ensuring good IEQ (Asim et al., 2022). HVAC systems should be inspected, cleaned, and maintained regularly to prevent the accumulation of dust, microbial growth, and other contaminants, and to ensure efficient system operation. Studies have shown that poor maintenance of HVAC systems can lead to poor indoor air quality and may increase the risk of respiratory illness (Asim et al., 2022; Au-Yong et al., 2014). Well-maintained HVAC systems contribute not only to thermal comfort but also to healthier indoor air and more reliable building performance.
4. **Use of low-VOC products**: VOCs are chemicals found in many indoor products, including paints, adhesives, cleaners, furnishings, and finishes. Exposure to high levels of VOCs can cause headaches, dizziness, irritation, and other health problems. The use of low-VOC products can significantly reduce indoor air pollution and improve IEQ (Piasecki et al., 2021; Zhong et al., 2017). Selecting healthier materials and products during building design, fit-out, and operation is therefore an important strategy for reducing exposure to indoor pollutants.

5. **Regular cleaning and disinfection**: Regular cleaning and disinfection of indoor spaces can help reduce pollutant levels and improve IEQ. These activities remove dust, dirt, biological contaminants, and other particles from indoor surfaces, thereby supporting cleaner indoor air and healthier spaces. Studies have shown that regular cleaning and disinfection can reduce allergen levels and improve health outcomes (Haverinen-Shaughnessy and Shaughnessy, 2015; Wu et al., 2007). However, cleaning practices should also be carefully managed so that the products used do not introduce unnecessary chemical pollutants into the indoor environment.
6. **Lighting**: Good lighting is important for occupant comfort, visual performance, and productivity. Daylighting can provide natural light, which has been shown to improve health, mood, and well-being. Artificial lighting should also be carefully designed to minimise glare, flicker, and visual discomfort, all of which may contribute to eye strain, headaches, and reduced performance (Galasiu and Veitch, 2006; Küller et al., 2007). A balanced approach that combines natural and artificial lighting appropriately can significantly improve the indoor environment.
7. **Noise control**: Excessive noise can lead to annoyance, stress, distraction, and even hearing-related problems. It is therefore important to design and manage buildings to reduce noise from external sources, such as traffic, and internal sources, such as HVAC systems and mechanical equipment. Sound-absorbing materials and acoustic design features, including baffles, diffusers, and appropriate interior finishes, can be used to improve indoor acoustic quality (Ma et al., 2018). Good acoustic control is essential for creating indoor spaces that support concentration, communication, rest, and occupant comfort.

Improving IEQ is essential for protecting health, supporting well-being, and creating productive indoor spaces. Strategies such as natural ventilation, proper air filtration, effective HVAC maintenance, low-VOC products, regular cleaning and disinfection, good lighting, and noise control can all contribute significantly to improved IEQ. When applied together, these strategies can reduce health risks and create healthier, safer, and more comfortable buildings.

5.5 Balancing Indoor Environmental Quality and Energy Usage

As a result of recent international legislation, policy developments, and professional guidelines, IEQ has become an important focus in both academia and industry. One of the most difficult issues in this area is balancing occupant health and comfort with the energy required to maintain them. Buildings consume energy to support ventilation, heating, cooling, and lighting, all of which are necessary for maintaining healthy and comfortable indoor environments. For this reason, building operation can often be understood as a cost-and-benefit dilemma in which energy consumption and IEQ must be carefully balanced (Geng, Lin and Zhu, 2020). These two factors remain among the most important indicators of operational building performance.

Past experience has often suggested that reducing energy use while simultaneously improving IEQ may be difficult and, in some cases, appear mutually exclusive (Roulet, 2001; Roulet et al., 2006). This perception has shaped many debates within building operations and environmental management. However, more recent policy directions have shown that the two goals should not be treated as incompatible. In Europe, for example, Directive (EU) 2018/844 highlights the need to improve occupant comfort and well-being alongside energy efficiency and broader building performance. This reflects a growing recognition that building performance should be judged not only by reduced energy demand but also by the quality of the indoor environment experienced by users.

Danza et al. (2020) noted that many previous studies, as well as common practice, have highlighted the difficulty of addressing IEQ and its individual components in an integrated way. The EPA (2000) also reported disputes between IEQ and energy efficiency, particularly where energy-saving strategies relied on reducing ventilation rates or loosely controlling conditions such as temperature and relative humidity. Such practices may save energy in the short term, but they can also weaken IEQ and compromise occupant well-being. This explains why balancing IEQ and energy use remains a central concern in sustainable building operation.

Even so, Seppęnen and Seppänen (2008) and Persily and Emmerich (2012) argued that the pursuit of both a healthy indoor environment and energy efficiency does not need to be mutually exclusive. It is possible to achieve both through appropriate design strategies, technologies, and management systems. Sustainable solutions such as natural lighting, natural ventilation, green energy systems, thermal recovery technologies, and advanced environmental monitoring and control systems can help create indoor environments that are both healthy and energy-efficient (Chwieduk, 2003; Chel and Kaushik, 2018; Shaik, 2015). In this way, the challenge is not simply whether both goals can be achieved, but how buildings are designed, operated, and managed to support them simultaneously.

After decades of strong international growth, green buildings have become an important trend in the building industry (Zuo and Zhao, 2014). In general, the IEQ of buildings has improved since the rise of research and practice in sustainable development and green building (Geng, Lin and Zhu, 2020). Within this broader movement, the concept of the Zero-Energy Building (ZEB) emerged as part of the sustainable building agenda. When designing a ZEB, a comprehensive and integrated strategy is required in order to achieve both high energy efficiency and a level of IEQ that demonstrates long-term, economically sustainable performance (Danza et al., 2020). This shows that balancing IEQ and energy use is not only possible but essential to the future of sustainable, high-performing buildings.

5.6 Health Effects of Poor Indoor Environmental Quality

IEQ is a major determinant of the health and well-being of building occupants. When indoor environmental conditions are poor, the effects extend far beyond simple discomfort. Poor IEQ has been associated with a wide range of adverse outcomes, including respiratory illness, cardiovascular problems, allergies, headaches, fatigue, reduced cognitive performance, and mental health challenges (Attia, 2018). As people spend a large proportion of their time indoors, the quality of indoor environments directly influences daily health, functioning, and quality of life. This makes IEQ an essential concern in the design, operation, and management of buildings.

One of the most widely reported effects of poor IEQ relates to respiratory health. Exposure to indoor pollutants such as formaldehyde, VOCs, and nitrogen dioxide (NO2) has been linked to higher rates of asthma, bronchitis, and other respiratory diseases (Dales et al., 2008). Poor indoor air quality has also been associated with wheezing, coughing, respiratory infections, and reduced lung function (Hulin et al., 2012). In addition, poor IEQ may also affect cardiovascular health. Exposure to fine particulate matter, especially PM2.5, has been associated with increased cardiovascular mortality and a greater risk of events such as heart attacks and strokes (Hayes et al., 2020). Noise exposure within the built environment has likewise been linked to hypertension and cardiovascular disease (Stansfeld and Matheson, 2003). These findings show that poor IEQ can undermine both respiratory and cardiovascular health over time.

Poor IEQ also affects human functioning in less visible but equally important ways. Conditions such as inadequate lighting, excessive noise, poor air quality, and uncomfortable temperatures have been linked to lower productivity, reduced concentration, and impaired cognitive performance (Al Horr et al., 2016a). Workers in buildings with poor IEQ have reported greater fatigue and lower productivity than those in healthier indoor environments (Sadick, Kpamma and Agyefi-Mensah, 2020). Reviews of the literature also show that lighting, noise, and thermal conditions significantly affect the cognitive performance and well-being of building users (Seppänen and Fisk, 2006). Poor IEQ has also been associated with mental health concerns such as stress, anxiety, and depression. By contrast, access to natural light and views of nature has been shown to improve recovery, mood, and cognitive functioning (Ulrich, 1984; Bedrosian and Nelson, 2017; Ma et al., 2018). These findings underscore that IEQ affects not only physical health but also emotional and psychological well-being.

Other common effects of poor IEQ include allergies, eye irritation, and headaches. Indoor air pollutants and allergens, such as dust mites, pet dander, mould, formaldehyde, and VOCs, can exacerbate symptoms such as rhinitis, conjunctivitis, sneezing, and itchy eyes (Richardson, Eick and Jones, 2005). Eye irritation has also been associated with high levels of particulate matter and VOC exposure (Derby et al., 2017). Headaches are another frequent complaint in indoor spaces with poor air quality, inadequate ventilation, or excessive noise, and have been linked to pollutants such as CO, nitrogen dioxide, and particulate matter (Reynolds et al., 2001; Tietjen et al., 2012; Ismail et al., 2022). Collectively, poor IEQ can have serious and wide-ranging effects on human health and well-being. For this reason, building occupants, designers, managers, and policymakers must prioritise strategies that support clean air, proper ventilation, acceptable temperature and humidity, and healthier indoor conditions.

5.7 Enhancing IEQ Management Using Digital Twin

IEQ is a critical aspect of building design, operation, and management. It encompasses factors such as indoor air quality, thermal comfort, lighting, and acoustics, all of which directly influence occupant health, well-being, productivity, and satisfaction. As people spend a significant proportion of their time indoors, maintaining high IEQ standards is essential for creating healthy and supportive built environments. In this context, DT, as a virtual replica of a physical asset or system, has emerged as a promising approach to enhancing IEQ management. Its value lies in its ability to support real-time monitoring, analysis, prediction, and optimisation of indoor environmental conditions (van Midden, 2022). Rather than relying only on periodic inspections or reactive interventions, DT allows building managers to engage with IEQ in a more continuous, data-driven, and intelligent manner.

One of the most important ways in which DT enhances IEQ management is through real-time monitoring and analysis. DT can be used to monitor and assess a wide range of IEQ variables, including indoor air quality, temperature, humidity, lighting conditions, and noise levels. By collecting and processing data from sensors and Internet of Things (IoT) devices installed throughout a building, DT can provide building managers with a more comprehensive understanding of the current indoor environment. This allows potential problems that may affect occupant comfort, health, and well-being to be identified more quickly and more accurately (Re Cecconi et al., 2021). In this regard, DT supports a stronger relationship between actual building conditions and the decisions taken to improve them.

DT also strengthens IEQ management through predictive and preventive maintenance. By combining machine learning techniques with historical and real-time data, DT can help predict equipment failure, system malfunction, and other technical issues that may negatively affect IEQ. This means potential problems can be identified before they occur and begin to affect building occupants. Building managers can then implement preventive maintenance actions, such as replacing worn components, cleaning filters, recalibrating sensors, or adjusting HVAC settings, in order to maintain optimal IEQ performance (Cai et al., 2023). This proactive approach reduces the likelihood of system breakdown and helps ensure that indoor conditions remain stable and healthy over time.

Another important contribution of DT is the optimisation of building systems that directly affect IEQ. DT can be used to simulate different operational scenarios and assess their effect on indoor environmental performance. This can apply to systems such as HVAC, lighting, shading, and other building services. By evaluating the relationship between energy efficiency, occupant comfort, and system performance, building managers are better positioned to identify strategies that maintain desirable IEQ conditions while also reducing energy consumption and related cost (Cai et al., 2023). In this way, DT supports more balanced and intelligent building operation, especially where environmental quality and energy performance must be considered together.

DT can also support performance benchmarking and continuous improvement in IEQ management. It enables the establishment of performance benchmarks for different IEQ factors and the comparison of actual building conditions against industry standards, best practices, or the performance of similar buildings. Through ongoing monitoring and analysis, building managers can identify areas of weakness, introduce targeted interventions, and assess whether those interventions lead to measurable improvement over time (van Midden, 2022). This creates a stronger basis for continuous improvement and makes IEQ management more transparent, measurable, and accountable.

The use of DT in IEQ management offers several important advantages. These include improved occupant comfort and well-being, better energy efficiency, more proactive maintenance, and stronger data-driven decision-making. By integrating DT into building management processes, managers can monitor, analyse, and optimise IEQ more effectively, thereby creating healthier, more comfortable indoor environments while reducing buildings' environmental impact. As technology advances and the demand for sustainable, occupant-centred buildings grows, DT is likely to play an increasingly important role in the future of IEQ management.

5.7.1 Benefits of the Digital Twin for IEQ Management

Implementing a DTS for IEQ management in any space offers several important benefits for building owners, operators, managers, and occupants. One of the most immediate benefits is real-time monitoring. The DTS enables the continuous monitoring of IEQ parameters within the environment, thereby providing building managers with up-to-date information on the quality of the indoor environment (Re Cecconi et al., 2021). This continuous visibility is important because IEQ conditions can change throughout the day in response to occupancy, weather, equipment use, and building system performance. With real-time information, managers are better positioned to detect changes early and respond more effectively. Another major benefit is enhanced occupant comfort and well-being. By enabling real-time monitoring, analysis, and optimisation of IEQ factors, DT can help maintain a more comfortable, healthy, and productive indoor environment. This can improve occupant satisfaction, reduce discomfort, and support better well-being (van Midden, 2022). Since building occupants spend long periods indoors, even small improvements in air quality, thermal comfort, lighting, and acoustics can make a meaningful difference to their daily experience and performance.

The DTS also supports proactive maintenance. Through continuous data collection and system analysis, it can identify patterns indicating potential issues or malfunctions within building systems that affect IEQ. This allows building managers to schedule maintenance in advance rather than reacting only after unexpected breakdowns occur (Cai et al., 2023). Such a proactive approach can reduce disruption, improve reliability, and help maintain more stable indoor environmental conditions over time. Another important benefit is improved decision-making. The DTS provides building managers with real-time data on IEQ conditions and energy use, thereby supporting more informed decisions about building operations, maintenance, and control strategies. Over time, this can improve operational efficiency and contribute to cost savings (Saini et al., 2021). Instead of relying on assumptions or delayed feedback, managers can act on current and evidence-based information.

The DTS also contributes to energy efficiency. By supporting the intelligent control of HVAC systems, lighting, and window blinds, the system can help optimise energy use in buildings. This can reduce unnecessary energy consumption and lower operating costs for the building owner, while still maintaining acceptable IEQ conditions (van Midden, 2022). In this sense, DT helps create a better balance between occupant comfort and energy performance. Finally, DT can enhance collaboration and communication. It can serve as a central platform for sharing information and insights among building stakeholders, including managers, engineers, architects, and occupants. This supports better coordination and improves communication around IEQ management initiatives.

5.7.2 Adapting Basic Digital Twin Requirements for IEQ Management

The basic requirements for deploying DT, as discussed earlier in Chapter 2, Section 2.4.1 can be effectively adapted to develop a DT-enabled IEQ management system. In the context of IEQ, these requirements do not change in principle, but they must be interpreted in relation to the specific needs of indoor environmental monitoring and control. A DT-enabled IEQ management system must therefore be structured to support continuous sensing, rapid analysis, intelligent response, and stable building performance. In this sense, the basic DT requirements become practical conditions for achieving healthier, more comfortable, and more efficiently managed indoor environments.

One of the most important requirements is **autonomy**. The system should be able to operate with minimal, not constant, human supervision so that IEQ conditions can be continuously monitored and managed. To achieve this, the system must be equipped with algorithms that can learn from environmental conditions and respond with corrective actions and automated control of physical systems. Closely related to this is **connectivity**. The system must be able to collect and transmit real-time data from sensors and IoT devices installed in a space. These sensors should be strategically positioned to monitor key IEQ parameters, while the system communicates with them via a reliable wireless network and stores the resulting data in a central database for access, interpretation, and analysis. **Real-time** capability is equally essential. The system should be able to collect, process, and analyse data immediately so that out-of-range conditions can be detected without delay and corrective actions can be recommended or implemented promptly.

Reliability is another critical requirement. A DT-enabled IEQ system must produce accurate, available, and consistent data so that decisions based on it are dependable. This requires routine equipment calibration and a system that remains functional with minimal interruption. **Interoperability** is also essential. The platform should be able to integrate with other systems and applications, exchange information smoothly, and use processed data to influence physical systems in buildings. In the same way, **data exchange and integration** must be robust enough to combine information from sensors, IoT devices, and related systems into a single usable, accessible environment. This gives building managers a more complete view of IEQ conditions and supports better decision-making.

The system must also support **resource optimisation**. It should be able to maintain acceptable IEQ conditions while reducing unnecessary energy use and other resource waste. Through continuous analysis, the system should identify where adjustments can improve both indoor conditions and sustainability outcomes. **Adaptability** is equally important. Since indoor environments are constantly changing, the system must adapt to evolving conditions and improve its behaviour over time. As more data are gathered, machine learning capabilities can further strengthen this adaptive function. Finally, **data security** remains fundamental. As the system collects and stores sensitive environmental and operational information, strong protection measures are needed to ensure confidentiality, restrict access, and prevent misuse. Together, these adapted requirements form the operational foundation of an effective DT-enabled IEQ management system.

5.8 Theoretical Underpinning of Digital Twin for IEQ Management

The integration of IEQ management with DT technology represents a forward-looking approach to improving the health, comfort, and productivity of building occupants. This integration brings together building science, environmental management, digital technologies, and human-centred performance thinking into a single, coordinated system. It reflects the growing need to manage indoor environments in ways that are not only more intelligent and data-driven but also more responsive to users' changing needs. In this context, theory plays an important role by providing the conceptual foundation for understanding how DT can be structured, accepted, applied, and evaluated within IEQ management.

A number of theoretical and conceptual frameworks help to explain this relationship. These frameworks provide essential insights into how digital representations of physical systems can be used to monitor, analyse, and improve indoor environmental conditions. They also help to clarify how technology can support decision-making, how users may respond to such systems, and how wider performance outcomes can be assessed. In the case of IEQ management, this is especially important because the subject is not purely technical. It also involves human perception, health, behavioural response, organisational practice, and long-term sustainability. A strong theoretical grounding is therefore necessary if DT is to be applied meaningfully and effectively in this area.

By reviewing relevant theoretical frameworks, it becomes easier to understand the multidisciplinary approach required for DT-enabled IEQ management. Such an approach must connect technological innovation with practical building management, occupant needs, and broader sustainability goals. It must also account for how information is generated, interpreted, and acted upon in the indoor environment. In this regard, the theoretical underpinning of DT for IEQ management provides more than just academic support. It helps establish the logic through which responsive, adaptive, and performance-oriented indoor environments can be achieved.

The sections that follow present the key theoretical perspectives that support this integration. These include the Technology Acceptance Model (TAM), the Indoor Environmental Quality–Health Performance Indicator (IEQ-HPI), and the Triple Bottom Line (TBL) approach. Together, these frameworks provide a useful basis for understanding the acceptance, performance, and broader value of DT for IEQ management.

5.8.1 Technology Acceptance Model

The TAM, originally proposed by Davis (1989), has made a major contribution to the understanding of user acceptance and usage behaviour in relation to information systems. In an increasingly digital age, understanding how people accept and use new technologies is important both in theory and in practice, especially when the goal is to design and implement systems that are effective, practical, and user-friendly. TAM offers a strong theoretical lens for explaining and predicting technology adoption by focusing on two central determinants: perceived usefulness and perceived ease of use (Davis, 1989). Over time, the model has proved valuable in many contexts, including business, education, and healthcare, and has become one of the most widely cited models in information systems.

TAM emerged from the Theory of Reasoned Action (TRA), proposed by Ajzen (1980). TRA explains that people's behaviour is driven by their behavioural intentions, which, in turn, are shaped by their attitudes towards the behaviour and subjective norms. Recognising that TRA did not fully capture the specific realities of technology adoption and usage, Davis (1989) introduced TAM as a more focused framework for understanding information system acceptance. TAM simplifies TRA by assuming that behavioural intention is the closest determinant of actual system use and that this intention is influenced mainly by two beliefs: perceived usefulness and perceived ease of use. In this model, perceived usefulness refers to the degree to which a person believes that using a particular system would enhance job performance, while perceived ease of use refers to the degree to which a person believes that using the system would require little effort (Davis, 1989). Together, these two constructs explain whether a user is likely to accept and use a technology.

The model has continued to evolve over time. Later versions, particularly TAM2 and TAM3, introduced additional variables such as subjective norms, output quality, result demonstrability, social influence processes, cognitive instrumental processes, and the role of experience and voluntariness (Venkatesh and Davis, 2000; Venkatesh and Bala, 2008). These additions improved the model's explanatory and predictive power by showing that technology acceptance is influenced not only by usefulness and ease of use but also by social and contextual conditions. Even with these developments, the original logic of TAM remains highly relevant because it continues to offer a clear and practical way of understanding why some technologies are accepted more readily than others.

In the context of DT for IEQ management, TAM provides a useful foundation for understanding how end users may adopt and use such systems in building management environments. A DT-enabled IEQ management system may offer strong technical capabilities, but its success also depends on whether users find it useful and easy to use. TAM therefore helps explain how building managers, facility personnel, and other stakeholders may respond to the system. It can support the prediction of user behaviour, guide successful implementation, and help refine the system's design by focusing on usefulness and usability. It can also help assess users' readiness to accept this more proactive and intelligent approach to IEQ management, thereby supporting more effective deployment and long-term adoption.

5.8.2 Indoor Environmental Quality–Health Performance Indicator

The IEQ-HPI is a theoretical model developed to support the systematic monitoring and management of IEQ. IEQ is a vital aspect of the built environment because it directly influences the health, comfort, and productivity of occupants (Allen et al., 2016). For this reason, IEQ has received considerable scholarly attention, with many studies emphasising its importance in creating healthy and supportive indoor environments (Bakó-Biró et al., 2012). IEQ covers several interrelated dimensions, including thermal comfort, air quality, lighting, and acoustics. These dimensions interact continuously to shape occupants' perception of indoor spaces and, in turn, influence their health, well-being, and performance (Frontczak and Wargocki, 2011). To monitor and manage these dimensions more effectively, the IEQ-HPI was proposed as a quantifiable, standardised measure to support the evaluation and comparison of indoor environments across different settings (Abboushi et al., 2022). In this regard, it serves as a valuable tool for designers, building operators, and policymakers who seek to promote healthier and more human-centred built environments.

The determination of IEQ-HPI is influenced by several factors. These include, among others, air quality, thermal comfort, acoustics, and lighting (Frontczak and Wargocki, 2011; Nimlyat, Isa and Gofwen, 2017). Air quality is one of its most fundamental components and is commonly assessed by the presence and concentrations of pollutants such as VOCs, carbon dioxide, and particulate matter, along with related factors such as humidity and ventilation rates (Hodgson, 2002; Nimlyat, Isa and Gofwen, 2017). Thermal comfort is another major factor, referring to the degree of satisfaction people feel with their thermal environment. It is affected by environmental conditions such as air temperature, radiant temperature, relative humidity, and air velocity, as well as personal factors such as clothing insulation and metabolic rate (Nimlyat, Isa and Gofwen, 2017; Suzianti et al., 2017). Acoustic conditions also contribute to IEQ-HPI, since indoor noise levels and sound quality significantly affect comfort, concentration, and productivity (Nimlyat, Isa and Gofwen, 2017). Lighting is equally important because the quality, quantity, and type of indoor lighting affect visual comfort, health, and psychological well-being, all of which shape the IEQ-HPI (Nimlyat, Isa and Gofwen, 2017; Suzianti et al., 2017).

In addition to these core dimensions, other factors may also influence IEQ-HPI. These include occupants' perceptions, the type of activities they perform, and their health status (Wei et al., 2020). This shows that IEQ-HPI is not only a technical indicator. It is also a human-centred measure because it reflects the relationship between indoor environmental conditions and users' lived experience. Its importance in building design and operation is therefore considerable. IEQ-HPI provides a measurable standard for evaluating the effects of indoor environments on occupant health, comfort, and productivity (Nimlyat, Isa and Gofwen, 2017). A well-managed indoor environment can enhance occupant wellness, satisfaction, and performance. For example, good air quality can reduce the spread of disease and lower the occurrence of allergies and asthma-related symptoms, while a comfortable thermal environment can support concentration and improve satisfaction (Allen et al., 2016).

Recent research has also shown that IEQ can significantly influence cognitive function. Allen et al. (2016) found that participants in environments with better IEQ performed better on cognitive function tests. This finding highlights the importance of integrating IEQ-HPI into the design, operation, and management of buildings. It also reinforces the idea that the indoor environment should be viewed not solely from the perspective of physical comfort but also from those of human health and intellectual performance. IEQ-HPI, therefore, provides more than a measurement tool. It provides a practical basis for making decisions that improve occupant well-being and support healthier building outcomes.

Within the broader context of DT for IEQ management, IEQ-HPI plays a particularly important role. It offers a standardised, quantifiable means of assessing key aspects of IEQ, including air quality, thermal comfort, acoustics, and lighting. When integrated with DT, IEQ-HPI can contribute to creating a dynamic and responsive system that not only monitors these conditions but also supports their proactive optimisation. In this way, integrating IEQ-HPI into a DT framework can help ensure compliance with health-centred performance goals while also enhancing occupant comfort, well-being, and cognitive performance. More importantly, it represents an important link between technological innovation and human-centred building management. Through this integration, DT for IEQ management can move beyond technical control and become a more meaningful tool for achieving healthier, more sustainable, and more occupant-focused indoor environments (Fisk, 2017).

5.8.3 Triple Bottom Line Approach

The TBL is a theoretical framework that broadens the focus of organisational responsibility beyond the traditional financial bottom line to include social and environmental considerations. Coined by John Elkington in 1994, the concept represents a shift towards a more balanced and comprehensive approach to measuring organisational performance (Elkington, 1997). Rather than focusing solely on profitability, the TBL approach encourages organisations to evaluate success across the full range of their impacts on people, the environment, and the economy (Jeurissen, 2000). In this way, TBL moves performance assessment beyond short-term financial outcomes and towards a wider understanding of sustainable value.

The TBL approach has become an important foundation in the field of sustainable business practice. By encouraging organisations to consider the social, environmental, and financial implications of their actions, it promotes a more holistic view of responsibility and performance (Savitz and Weber, 2006). This perspective aligns with the growing global emphasis on sustainability, responsible management, and long-term resilience. The TBL framework is commonly explained through its three interrelated pillars: people, planet, and profit. These pillars represent the core dimensions through which sustainability can be understood and applied in practice.

The people dimension focuses on social equity, fairness, and the well-being of stakeholders. It includes concerns such as labour conditions, human rights, community development, and social engagement (Dyllick and Hockerts, 2002). The planet dimension addresses environmental responsibility. It emphasises reducing negative environmental impacts, responsible use of natural resources, energy efficiency, climate change mitigation, conservation, and sustainable sourcing (Hart, Milstein and Caggiano, 2003). The profit dimension refers to economic sustainability. It highlights the need for organisations to remain financially viable while also acting responsibly towards society and the environment. This includes long-term financial stability, ethical economic practice, and recognition that financial performance is closely connected to social and environmental outcomes (Slaper and Hall, 2011).

The TBL framework also aligns closely with international efforts to achieve sustainable development, including the United Nations Sustainable Development Goals (SDGs). The three pillars of TBL reflect the economic, social, and environmental dimensions of sustainability that are echoed across the 17 SDGs (Sachs, 2015). By following a balanced approach, organisations that embrace

TBL contribute to wider goals such as environmental protection, social equity, and improved human well-being (Hopwood, Mellor and O'Brien, 2005). This makes TBL especially relevant in fields where human health, environmental performance, and operational efficiency must be considered together.

In the context of IEQ management through DT, the TBL approach provides a useful framework for understanding how technological systems can support sustainable outcomes. From the social perspective, IEQ directly affects occupant health, comfort, satisfaction, and productivity. A DT can support a more proactive and responsive approach to managing air quality, temperature, lighting, and other indoor environmental conditions, thereby helping to create healthier, more supportive indoor environments. From the environmental perspective, DT can support real-time monitoring and control of energy use, resource optimisation, and waste reduction. These functions are important for improving environmental performance and reducing the ecological footprint of buildings. From the economic perspective, DT can help reduce operational costs through better system optimisation, more efficient maintenance, and improved use of resources. At the same time, improved occupant well-being and productivity can also generate wider economic benefits at the organisational level.

In essence, the TBL framework provides a comprehensive lens for understanding the relationship between IEQ management, DT, and sustainability. It helps to connect the technical capabilities of DT with broader social, environmental, and economic goals. In doing so, it supports a more balanced and meaningful approach to IEQ management, one that benefits building users, organisations, and the environment at the same time.

5.9 Summary

This chapter focused on the role of DT in managing IEQ and showed that IEQ remains a critical concern in creating healthy, comfortable, and productive indoor spaces. The discussion began by explaining the meaning and major components of IEQ, particularly air quality, thermal comfort, lighting, and acoustic quality. It also showed that human perception of IEQ extends beyond these core technical factors and is influenced by broader physical, sensory, and psychological conditions within indoor environments.

The chapter further explored practical strategies for improving IEQ, including ventilation, air filtration, HVAC maintenance, low-emission materials, cleaning practices, lighting quality, and noise control. It also discussed the long-standing challenge of balancing IEQ with energy usage, showing that healthy indoor conditions and energy efficiency should not be treated as opposing goals. In addition, the chapter highlighted the health effects of poor IEQ, demonstrating that inadequate indoor conditions can impair respiratory, cardiovascular, and cognitive health, as well as emotional well-being and overall comfort.

A major focus of the chapter was the use of DT to enhance IEQ management. The discussion showed that DT can support real-time monitoring, predictive maintenance, system optimisation, performance benchmarking, and more informed decision-making regarding indoor environmental conditions. The benefits of a DTS for IEQ management were also outlined, together with the adaptation of basic DT requirements to suit IEQ applications. Finally, the chapter considered the theoretical foundations of this integration through the TAM, the IEQ-HP Indicator, and the TBL approach. Ultimately, the chapter showed that DT offers a strong, forward-looking pathway to more intelligent, proactive, and human-centred IEQ management.

References

Abboushi, B. et al. (2022) 'A review of the use of wearables in indoor environmental quality studies and an evaluation of data accessibility from a wearable device', *Frontiers in Built Environment*, 8, p. 787289. doi:10.3389/FBUIL.2022.787289

Agbo, K.E. et al. (2021) 'A review on ambient and indoor air pollution status in Africa', *Atmospheric Pollution Research*, 12(2), pp. 243–260. doi:10.1016/J.APR.2020.11.006

Ajzen, I. (1980) *Understanding attitudes and predicting social behavior*. Englewood Cliffs, NJ: Prentice-Hall.

Al Horr, Y. et al. (2016a) 'Impact of indoor environmental quality on occupant well-being and comfort: A review of the literature', *International Journal of Sustainable Built Environment*, 5(1), pp. 1–11. doi:10.1016/J.IJSBE.2016.03.006

Al Horr, Y. et al. (2016b) 'Occupant productivity and office indoor environment quality: A review of the literature', *Building and Environment*, 105, pp. 369–389. doi:10.1016/J.BUILDENV.2016.06.001

Allen, J.G. et al. (2016) 'Associations of cognitive function scores with carbon dioxide, ventilation, and volatile organic compound exposures in office workers: A controlled exposure study of green and conventional office environments', *Environmental Health Perspectives*, 124(6), pp. 805–812. doi:10.1289/EHP.1510037

Al-Sulaihi, I. et al. (2015) 'Assessing indoor environmental quality of educational buildings using BIM', *Journal of Environmental Science and Engineering B*, 4(8). doi:10.17265/2162-5263/2015.08.006

American Society of Heating, Refrigerating, and Air-Conditioning Engineers (ASHRAE) (2020) *Standard 55 – Thermal environmental conditions for human occupancy*. Available at: https://www.ashrae.org/technical-resources/bookstore/standard-55-thermal-environmental-conditions-for-human-occupancy (Accessed: 6 April 2023).

Asim, N. et al. (2022) 'Sustainability of heating, ventilation and air-conditioning (HVAC) systems in buildings—An overview', *International Journal of Environmental Research and Public Health*, 19(2). doi:10.3390/IJERPH19021016

Attia, S. (2018) 'Occupants well-being and indoor environmental quality', *Net Zero Energy Buildings*, pp. 117–153. doi:10.1016/B978-0-12-812461-1.00005-8

Au-Yong, C.P., Ali, A.S. and Ahmad, F. (2014) 'Improving occupants' satisfaction with effective maintenance management of HVAC system in office buildings', *Automation in Construction*, 43, pp. 31–37. doi:10.1016/J.AUTCON.2014.03.013

Bakó-Biró, Z. et al. (2012) 'Ventilation rates in schools and pupils' performance', *Building and Environment*, 48(1), pp. 215–223. doi:10.1016/J.BUILDENV.2011.08.018

Bedrosian, T.A. and Nelson, R.J. (2017) 'Timing of light exposure affects mood and brain circuits', *Translational Psychiatry*, 7(1). doi:10.1038/TP.2016.262

Cai, J. et al. (2023) 'Digital twin for healthy indoor environment: A vision for the post-pandemic era', *Frontiers of Engineering Management*, pp. 1–19. doi:10.1007/S42524-022-0244-Y

Chel, A. and Kaushik, G. (2018) 'Renewable energy technologies for sustainable development of energy efficient building', *Alexandria Engineering Journal*, 57(2), pp. 655–669. doi:10.1016/J.AEJ.2017.02.027

Chwieduk, D. (2003) 'Towards sustainable-energy buildings', *Applied Energy*, 76(1–3), pp. 211–217. doi:10.1016/S0306-2619(03)00059-X

Daisey, J.M., Angell, W.J. and Apte, M.G. (2003) 'Indoor air quality, ventilation and health symptoms in schools: An analysis of existing information', *Indoor Air*, 13(1), pp. 53–64. doi:10.1034/J.1600-0668.2003.00153.X

Dales, R. et al. (2008) 'Public health: Quality of indoor residential air and health', *CMAJ*, 179(2), pp. 147–152. doi:10.1503/CMAJ.070359

Danza, L. et al. (2020) 'A weighting procedure to analyse the indoor environmental quality of a zero-energy building', *Building and Environment*, 183, p. 107155. doi:10.1016/j.buildenv.2020.107155

Davis, F.D. (1989) 'Perceived usefulness, perceived ease of use, and user acceptance of information technology', *MIS Quarterly: Management Information Systems*, 13(3), pp. 319–339. doi:10.2307/249008

Derby, M.M. et al. (2017) 'Update of the scientific evidence for specifying lower limit relative humidity levels for comfort, health, and indoor environmental quality in occupied spaces (RP-1630)', *Science and Technology for the Built Environment*, 23(1), pp. 30–45. doi:10.1080/23744731.2016.1206430

Dykes, C. and Baird, G. (2013) 'A review of questionnaire-based methods used for assessing and benchmarking indoor environmental quality', *Intelligent Buildings International*, 5(3), pp. 135–149. doi:10.1080/17508975.2013.783457

Dyllick, T. and Hockerts, K. (2002) 'Beyond the business case for corporate sustainability', *Business Strategy and the Environment*, 11(2), pp. 130–141. doi:10.1002/BSE.323

Elkington, J. (1997) *Cannibals with forks: The triple bottom line of 21st century business*. Oxford: Capstone Publishing.

Environmental Protection Agency (2022a) *Air cleaners and air filters in the home*. United States Environmental Protection Agency. Available at: https://www.epa.gov/indoor-air-quality-iaq/air-cleaners-and-air-filters-home (Accessed: 6 April 2023).

Environmental Protection Agency (2022b) *Introduction to indoor air quality*. United States Environmental Protection Agency. Available at: https://www.epa.gov/indoor-air-quality-iaq/introduction-indoor-air-quality (Accessed: 6 April 2023).

Fantozzi, F. and Rocca, M. (2020) 'An extensive collection of evaluation indicators to assess occupants' health and comfort in indoor environment' *Atmosphere*, 11(1), p. 90. doi:10.3390/ATMOS11010090

Fisk, W.J. (2017) 'The ventilation problem in schools: Literature review', *Indoor Air*, 27(6), pp. 1039–1051. doi:10.1111/INA.12403

Frontczak, M. and Wargocki, P. (2011) 'Literature survey on how different factors influence human comfort in indoor environments', *Building and Environment*, 46(4), pp. 922–937. doi:10.1016/j.buildenv.2010.10.021

Galasiu, A.D. and Veitch, J.A. (2006) 'Occupant preferences and satisfaction with the luminous environment and control systems in daylit offices: A literature review', *Energy and Buildings*, 38(7), pp. 728–742. doi:10.1016/J.ENBUILD.2006.03.001

Geng, Y., Lin, B. and Zhu, Y. (2020) 'Comparative study on indoor environmental quality of green office buildings with different levels of energy use intensity', *Building and Environment*, 168, p. 106482. doi:10.1016/j.buildenv.2019.106482

Goyal, R. and Khare, M. (2011) 'Indoor air quality modeling for PM10, PM2.5, and PM1.0 in naturally ventilated classrooms of an urban Indian school building', *Environmental Monitoring and Assessment*, 176(1–4), pp. 501–516. doi:10.1007/S10661-010-1600-7

Hart, S.L., Milstein, M.B. and Caggiano, J. (2003) 'Creating sustainable value', *Academy of Management Executive*, 17(2), pp. 56–69. doi:10.5465/AME.2003.10025194

Haverinen-Shaughnessy, U. and Shaughnessy, R.J. (2015) 'Effects of classroom ventilation rate and temperature on students' test scores', *PLoS ONE*, 10(8). doi:10.1371/JOURNAL.PONE.0136165

Hayes, R.B. et al. (2020) 'PM2.5 air pollution and cause-specific cardiovascular disease mortality', *International Journal of Epidemiology*, 49(1), p. 25. doi:10.1093/IJE/DYZ114

Heinzerling, D. et al. (2013) 'Indoor environmental quality assessment models: A literature review and a proposed weighting and classification scheme', *Building and Environment*, 70, pp. 210–222. doi:10.1016/J.BUILDENV.2013.08.027

Hodgson, M. (2002) 'Indoor environmental exposures and symptoms', *Environmental Health Perspectives*, 110(Suppl 4), p. 663. doi:10.1289/EHP.02110S4663

Holm, D. and Engelbrecht, F.A. (2005) 'Practical choice of thermal comfort scale and range in naturally ventilated buildings in South Africa', *Journal of the South African Institution of Civil Engineering*, 47(2), pp. 9–14.

Hopwood, B., Mellor, M. and O'Brien, G. (2005) 'Sustainable development: mapping different approaches', *Sustainable Development*, 13(1), pp. 38–52. doi:10.1002/SD.244

Hulin, M. et al. (2012) 'Respiratory health and indoor air pollutants based on quantitative exposure assessments', *European Respiratory Journal*, 40(4), pp. 1033–1045. doi:10.1183/09031936.00159011

Humphreys, M.A. (2007) 'Quantifying occupant comfort: are combined indices of the indoor environment practicable?', *Building Research & Information*, 33(4), pp. 317–325. doi:10.1080/09613210500161950

Ismail, S.A. et al. (2022) 'Indoor air quality level influence sick building syndrome among occupants in educational buildings', *International Journal of Public Health Science (IJPHS)*, 11(2), pp. 503–517. doi:10.11591/IJPHS.V11I2.21125

Jeurissen, R. (2000) 'John Elkington, Cannibals with forks: The triple bottom line of 21st century business', *Journal of Business Ethics*, 23(2), pp. 229–231. doi:10.1023/A:1006129603978

Khajenasiri, F., Zamanian, A. and Zamanian, Z. (2016) 'The effect of exposure to high noise levels on the performance and rate of error in manual activities', *Electronic Physician*, 8(3), p. 2088. doi:10.19082/2088

Khorshid, R.F. et al. (2021) 'The effect of fluorescent light on anxiety patients', *Cureus*, 13(2). doi:10.7759/CUREUS.13436

Küller, R. et al. (2007) 'The impact of light and colour on psychological mood: a cross-cultural study of indoor work environments', *Ergonomics*, 49(14), pp. 1496–1507. doi:10.1080/00140130600858142

Lan, L. et al. (2011) 'Effects of thermal discomfort in an office on perceived air quality, SBS symptoms, physiological responses, and human performance', *Indoor Air*, 21(5), pp. 376–390. doi:10.1111/J.1600-0668.2011.00714.X

Levin, H. (2006) 'Editorial indoor environmental quality', *Building Research and Information*, pp. 433–436. doi:10.1080/09613210600931302

Ma, J. et al. (2018) 'A multilevel analysis of perceived noise pollution, geographic contexts and mental health in Beijing', *International Journal of Environmental Research and Public Health*, 15(7). doi:10.3390/IJERPH15071479

Marino, C., Nucara, A. and Pietrafesa, M. (2012) 'Proposal of comfort classification indexes suitable for both single environments and whole buildings', *Building and Environment*, 57, pp. 58–67. doi:10.1016/J.BUILDENV.2012.04.012

Mendell, M.J. et al. (2013) 'Association of classroom ventilation with reduced illness absence: a prospective study in California elementary schools', *Indoor Air*, 23(6), pp. 515–528. doi:10.1111/INA.12042

Ncube, M. and Riffat, S. (2012) 'Developing an indoor environment quality tool for assessment of mechanically ventilated office buildings in the UK – A preliminary study', *Building and Environment*, 53, pp. 26–33. doi:10.1016/J.BUILDENV.2012.01.003

Nimlyat, P.S. (2018) 'Indoor environmental quality performance and occupants' satisfaction [IEQPOS] as assessment criteria for green healthcare building rating', *Building and Environment*, 144, pp. 598–610. doi:10.1016/J.BUILDENV.2018.09.003

Nimlyat, P.S., Isa, A.A. and Gofwen, N.C. (2017) 'Performance indicators of indoor environmental quality (IEQ) assessment in hospital buildings: A confirmatory factor analysis (CFA) approach', *ATBU Journal of Environmental Technology*, 10(1), pp. 139–159.

Persily, A. K. and Emmerich, S. J. (2012) 'Indoor air quality in sustainable, energy efficient buildings', *HVAC&R Research*, 18(1–2), 4–20. doi:10.1080/10789669.2011.592106

Piasecki, M., Kostyrko, K.B. and Goljan, A. (2021) 'The ability to control VOC emissions from multilayer building materials', *Applied Sciences*, 11(11), p. 4806. doi:10.3390/APP11114806

Re Cecconi, F. et al. (2021) *Soft Digital Twin for IEQ enabling the COVID risk mitigation in educational spaces*. Available at: https://re.public.polimi.it/handle/11311/1187733 (Accessed: 30 April 2023).

Residovic, C. (2017) 'The new NABERS indoor environment tool – The next frontier for Australian buildings', *Procedia Engineering*, 180, pp. 303–310. doi:10.1016/J.PROENG.2017.04.189

Reynolds, S.J. et al. (2001) 'Indoor environmental quality in six commercial office buildings in the midwest United States', *Applied Occupational and Environmental Hygiene*, 16(11), pp. 1065–1077. doi:10.1080/104732201753214170

Richardson, G., Eick, S. and Jones, R. (2005) 'How is the indoor environment related to asthma? Literature review', *Journal of Advanced Nursing*, 52(3), pp. 328–339. doi:10.1111/j.1365-2648.2005.03591.x

Roulet, C. A. (2001) 'Indoor environment quality in buildings and its impact on outdoor environment', *Energy and Buildings*, 33, pp. 183–191. doi:10.1016/S0378-7788(00)00080-3

Roulet, C. A., Johner, N., Foradini, F., Bluyssen, P., Cox, C., De Oliveira Fernandes, E., Müller, B. and Aizlewood, C. (2006).'Perceived health and comfort in relation to energy use and building characteristics', *Building Research & Information*, 34(5), 467–474. doi:10.1080/09613210600822279

Sachs, J.D. (2015) *The age of sustainable development*. New York: Columbia University Press.

Sadick, A.M., Kpamma, Z.E. and Agyefi-Mensah, S. (2020) 'Impact of indoor environmental quality on job satisfaction and self-reported productivity of university employees in a tropical African climate', *Building and Environment*, 181. doi:10.1016/J.BUILDENV.2020.107102

Saini, J., Dutta, M. and Marques, G. (2021) 'Sensors for indoor air quality monitoring and assessment through Internet of Things: A systematic review', *Environmental Monitoring and Assessment*, 193(2), pp. 1–32. doi:10.1007/S10661-020-08781-6

Sakhare, V.V. and Ralegaonkar, R.V. (2014) 'Indoor environmental quality: Review of parameters and assessment models', *Architectural Science Review*, pp. 147–154. doi:10.1080/00038628.2013.862609

Savitz, A.W. and Weber, K. (2006) *The triple bottom line: how today's best-run companies are achieving economic, social, and environmental success - And how you can too*. Jossey-Bass.

Seppänen, O.A. and Fisk, W. (2006) 'Some quantitative relations between indoor environmental quality and work performance or health', *HVAC&R Research*, 12(4), pp. 957–973. doi:10.1080/10789669.2006.10391446

Seppęnen, O. (2008) 'Ventilation strategies for good indoor air quality and energy efficiency', *International Journal of Ventilation*, 6(4), pp. 297–306.

Shaik, M. (2015) *How to bridge Africa's infrastructure gap*. World Economic Forum. Available at: https://www.weforum.org/agenda/2015/06/how-to-bridge-africas-infrastructure-gap/ (Accessed: 8 September 2019).

Slaper, T.F. and Hall, T.J. (2011) 'The triple bottom line: What is it and how does it work?', *Indiana Business Review*, 86(1), pp. 4–8.

Stansfeld, S.A. and Matheson, M.P. (2003) 'Noise pollution: Non-auditory effects on health', *British Medical Bulletin*, 68, pp. 243–257. doi:10.1093/BMB/LDG033

Suzianti, A., Suradi, J., Nurtjahyo, B. and Brakumara, Y. (2017) 'Analysis of the indoor environmental quality's impacts on human performance', *International Journal of Mechanical Engineering and Robotics Research*, 6(2), pp. 164–169. doi:10.18178/IJMERR.6.2.164-169

Tietjen, A.K. et al. (2012) 'Headaches and indoor environmental quality among office workers', *Occupational Medicine*, 62(7), pp. 533–540. doi:10.1093/OCCMED/KQS124

Turunen, M. et al. (2014) 'Indoor environmental quality in school buildings, and the health and wellbeing of students', *International Journal of Hygiene and Environmental Health*, 217(7), pp. 733–739. doi:10.1016/j.ijheh.2014.03.002

Ulrich, R.S. (1984) 'View through a window may influence recovery from surgery', *Science*, 224(4647), pp. 420–421. doi:10.1126/SCIENCE.6143402

US Environmental Protection Agency (2016) *Green building*. Available at: https://archive.epa.gov/greenbuilding/web/html/about.html

van den Berg, A.E., Wesselius, J.E., Maas, J. and Tanja-Dijkstra, K. (2016) 'Green walls for a restorative classroom environment: A controlled evaluation study'. *Environment and Behavior*, 49(7), pp. 791–813. doi:10.1177/0013916516667976

van Midden, J.N.A. (2022) Linked data-based Digital Twins for increasing the indoor environmental quality (IEQ) of individuals working from home [Master's dissertation]. Eindhoven University of Technology. Available at: https://research.tue.nl/en/studentTheses/linked-data-based-digital-twins-for-increasing-the-indoor-environ (Accessed: 30 April 2023).

Venkatesh, V. and Bala, H. (2008) 'Technology acceptance model 3 and a research agenda on interventions', *Decision Sciences*, 39(2), pp. 273–315. doi:10.1111/J.1540-5915.2008.00192.X

Venkatesh, V. and Davis, F.D. (2000) 'Theoretical extension of the technology acceptance model: Four longitudinal field studies', *Management Science*, 46(2), pp. 186–204. doi:10.1287/mnsc.46.2.186.11926

Wålinder, R. et al. (2007) 'Physiological and psychological stress reactions in relation to classroom noise', *Scandinavian Journal of Work, Environment & Health*, 33(4), pp. 260–266. doi:10.5271/sjweh.1141

Wei, W. et al. (2020) 'Review of parameters used to assess the quality of the indoor environment in Green Building certification schemes for offices and hotels', *Energy and Buildings*, 209, p. 109683. doi:10.1016/J.ENBUILD.2019.109683

Wong, L.T., Mui, K.W. and Tsang, T.W. (2018) 'An open acceptance model for indoor environmental quality (IEQ)', *Building and Environment*, 142, pp. 371–378. doi:10.1016/j.buildenv.2018.06.031

World Health Organisation (WHO) (2010) *WHO guidelines for indoor air quality: Selected pollutants*. Copenhagen, Denmark: World Health Organisation. Available at: https://www.who.int/publications/i/item/9789289002134 (Accessed: 6 February 2022).

World Health Organisation (WHO) (2022) 'Ambient (outdoor) air pollution'. World Health Organisation Fact Sheets. Available at: https://www.who.int/news-room/fact-sheets/detail/ambient-(outdoor)-air-quality-and-health (Accessed: 6 April 2023).

Wu, F. et al. (2007) 'Improving Indoor Environmental Quality for Public Health: Impediments and Policy Recommendations', *Environmental Health Perspectives*, 115(6), p. 953. doi:10.1289/EHP.8986

Yang, W. and Moon, H.J. (2019) 'Combined effects of acoustic, thermal, and illumination conditions on the comfort of discrete senses and overall indoor environment', *Building and Environment*, 148, pp. 623–633. doi:10.1016/J.BUILDENV.2018.11.040

Yin, J. et al. (2018) 'Physiological and cognitive performance of exposure to biophilic indoor environment', *Building and Environment*, 132, pp. 255–262. doi:10.1016/J.BUILDENV.2018.01.006

Zhong, L., Su, F.C. and Batterman, S. (2017) 'Volatile organic compounds (VOCs) in conventional and high performance school buildings in the U.S.', *International Journal of Environmental Research and Public Health*, 14(1), p. 100. doi:10.3390/IJERPH14010100

Zuo, J. and Zhao, Z.Y. (2014) 'Green building research–current status and future agenda: A review', *Renewable and Sustainable Energy Reviews*, 30, pp. 271–281. doi:10.1016/J.RSER.2013.10.021

6 Digital Twin Model for Indoor Environmental Quality Management

6.1 Introduction

Managing indoor environmental quality (IEQ) requires more than occasional inspections and reactive measures. Indoor environments are constantly changing due to variations in occupancy, ventilation, temperature, lighting, equipment use, and user behaviour. As a result, successful IEQ management increasingly relies on systems that can continuously monitor conditions, interpret incoming data, detect issues early, and enable prompt responses. In this context, the Digital Twin System (DTS) offers a practical and intelligent solution by linking the physical indoor environment to a digital platform that enables real-time monitoring, analysis, and intervention. Consequently, the DTS delivers a more responsive and data-driven approach to IEQ management than many traditional building control methods (Re Cecconi et al., 2021; Cai et al., 2023; van Midden, 2022).

The DTS framework for IEQ management comprises several key components. These components work together to form a structured pathway for sensing, interpreting, evaluating, and correcting indoor environmental conditions. The framework includes the necessary data and devices, acceptable ranges and thresholds, connection to the digital system, digital rendering and display, issue identification, selection of corrective measures, and implementation of those measures. Each component has a specific role, yet they are also closely interdependent. Collectively, they create an integrated management system that allows for more effective monitoring and control of the indoor environment.

Figure 6.1 summarises the framework. The figure presents the DTS not as a single, isolated tool but as a connected system comprising multiple stages that support one another. The framework begins by identifying the data and devices required for monitoring. It then progresses through the setting of acceptable ranges and thresholds, the connection of the physical space to the digital system, and the rendering of collected information in a form that users can understand and act upon. Thereafter, the framework supports the identification of issues, the selection of corrective measures, and their implementation within the physical environment.

DOI: 10.1201/9781003743330-6

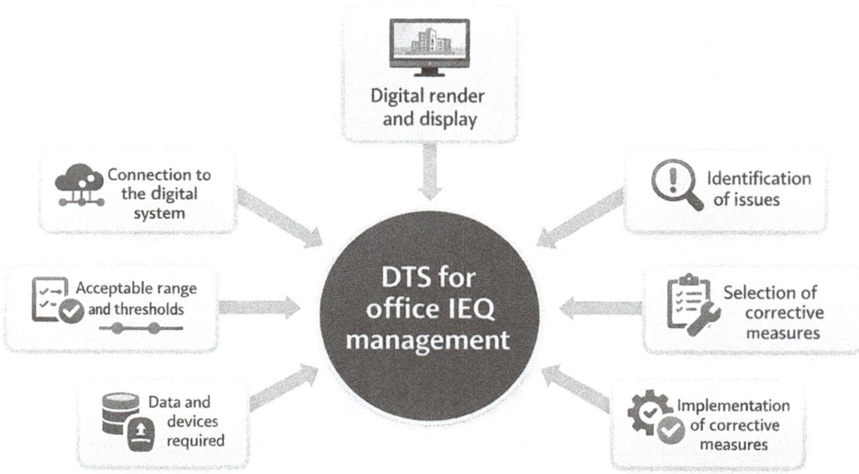

Figure 6.1 Framework of DTS for IEQ management.

By utilising the DTS framework, indoor environmental conditions can be monitored and maintained more efficiently, leading to improved occupant health, comfort, and productivity. The framework also supports faster response times, stronger decision-making, and a more proactive approach to IEQ management in indoor environments. These components are explored in detail in the sections that follow, thereby providing a structured basis for understanding the logic and practical application of the DTS framework for IEQ management.

6.2 Framework Component 1: Data and Devices Required

The first step in the framework for developing the DTS is identifying the data and devices required by the system. This step is fundamental because it determines the informational base upon which the entire system will operate. At this stage, the focus is on identifying the specific parameters or metrics relevant to the monitored indoor environment. The correct selection of data is crucial because such data forms the basis for monitoring, analysis, issue detection, and decision-making. If the wrong data are selected or important variables are omitted, the performance of the entire DTS will be compromised from the outset.

To determine which data should be collected, there must first be a clear understanding of the objectives of the monitoring system and the nature of the environment being monitored. In the context of IEQ management, this means understanding which indoor environmental conditions to monitor, why they matter, and how they relate to comfort, health, and operational performance. This understanding then makes it possible to determine the exact parameters to monitor and the frequency at which data should be collected. In other words, the purpose of the system shapes the nature, volume, and timing of the required data.

Once the required data has been identified, attention must be given to the devices that will collect, transmit, and, where necessary, respond to it. In this framework, the devices can generally be grouped into **sensors** and **actuators**. Sensors collect the required data from the physical environment, while actuators automate the actions of relevant physical features within the room or building. Figure 6.2 summarises this first framework component by outlining the main issues to consider regarding the required data and devices.

The first consideration is the **purpose of data collection**. This involves understanding why the data are being collected in the first place. The purpose of the collection usually determines the type of data needed and helps define the focus of the monitoring process (Khajavi et al., 2019; Singh et al., 2022). Without clarity of purpose, data collection can become excessive, unfocused, or insufficient for the intended management objectives.

The second consideration is the **type of data**. At this stage, it is necessary to understand what data are being collected and which measurements are required. This is important because the acceptable range and threshold values vary by parameter type. A correct understanding of the data type is therefore essential for setting suitable monitoring conditions and performance expectations (Kousi et al., 2019; Qi et al., 2019).

The third consideration is the **devices required**. Based on the purpose and type of data collection, the necessary devices can then be selected. As noted earlier, these devices generally fall into the categories of sensors and actuators. Sensors gather the necessary information from the physical environment, while actuators enable the system to automate specific responses or actions that affect physical features within the room or the monitored space (Zhu, Liu and Xu, 2019; Bende and Gochhait, 2020). The choice of devices must therefore match both the monitoring purpose and the intended system response.

Figure 6.2 Framework component 1: Data and devices required.

The fourth consideration is **data sensitivity**. It is important to carefully consider how sensitive the data may be, as this affects privacy, confidentiality, and system security. In some cases, the information being collected may relate not only to environmental conditions but also to the context in which it is gathered and used. This makes data sensitivity an important issue in the design of the DTS (Damjanovic-Behrendt, 2018; Murphy et al., 2020).

The fifth consideration is **data format**. The format in which data is collected and stored is important because it affects how easily the data can be processed, interpreted, and exchanged within the system. The data format should therefore be readable, accessible, and compatible with the digital system that will analyse and use it (Botkina et al., 2018; Haag and Anderl, 2019).

The sixth consideration is the **data sampling rate**. This refers to how often data are collected. The sampling rate must be chosen carefully because it affects the system's responsiveness, accuracy, and usefulness. The appropriate rate depends on the purpose of the data collection and on the level of precision required for the monitored parameter (Ganguli and Adhikari, 2020). If the rate is too low, important changes may be missed. If it is too high, the system may produce more data than is necessary or efficient to handle.

By carefully considering these conditions, the data and devices required for the DTS can be identified more accurately. This creates a strong, reliable foundation for the DTS and supports effective monitoring and management of IEQ in the indoor environment.

6.3 Framework Component 2: Acceptable Range and Thresholds

Once the data to be collected has been identified, the next step is to determine the acceptable range and threshold for each IEQ parameter. This step involves defining the normal or expected range for each parameter and setting threshold values for abnormal conditions that may indicate a problem. For effective IEQ management, acceptable ranges and thresholds are critical because they allow the DTS to identify issues in a timely and accurate manner. By clearly establishing these values, the system can quickly detect deviations from normal conditions and support prompt corrective action. Figure 6.3 summarises this second framework component, which focuses on acceptable ranges and thresholds.

The first issue to consider is **industry standards**. Relevant industry standards, guidelines, and regulations that apply to the data being collected must be considered, as they help determine what ranges are acceptable and which values should be regarded as critical thresholds (Lydon et al., 2019; Lu et al., 2020). In the context of IEQ, this is especially important because acceptable conditions for temperature, air quality, humidity, lighting, and acoustics are often shaped by recognised performance standards and health-related recommendations. These standards provide a technical and professional basis for setting operational limits within the DTS.

The second issue is **environmental consideration**. The specific environment in which the data are collected must be considered, as it can significantly affect

Figure 6.3 Framework component 2: Acceptable range and threshold.

the acceptable range and threshold values. Different indoor environments have different functional demands and user expectations. For example, the acceptable temperature range in a food storage facility will differ from that of an office, a classroom, or a healthcare facility (Uhlemann et al., 2017b; Haag and Anderl, 2018). This means that threshold setting cannot be treated as universal. It must reflect the actual context, function, and purpose of the environment being monitored.

The third issue is the **sensor accuracy**. The sensors used to collect environmental data must be accurate, reliable, and efficient. This is important because inaccurate sensors may produce misleading readings, which would affect the quality of the thresholds and the decisions based on them. Sensor accuracy can be supported through proper calibration, routine checks, and good maintenance practices (Tao et al., 2018; BSI Group, 2019). In other words, the value of a threshold depends not only on how it is defined, but also on the quality of the data used to compare against it.

The fourth issue is **adaptability**. The acceptable range and threshold values should be able to adjust to changing conditions in the monitored environment. These changes may arise from industry-specific requirements, building functions, seasonal variations, or even user-related IEQ preferences and comfort expectations. In this sense, adaptability is important because it allows the DTS to remain responsive to real-world conditions rather than relying on rigid, fixed assumptions (Vathoopan et al., 2018; Woodhead, Stephenson and Morrey, 2018). A system that can adapt its thresholds more intelligently is more likely to support meaningful and occupant-sensitive IEQ management.

By carefully considering these conditions, the acceptable range and threshold values for the collected data can be established more effectively. This creates a stronger basis for detecting abnormalities, interpreting environmental conditions, and supporting timely intervention within the DTS.

6.4 Framework Component 3: Connection to the Digital System

The next step in the framework involves establishing the connection to the digital system. At this stage, the DTS functions as the virtual representation of the physical room and serves as the platform through which data are collected, received, analysed, and interpreted in real time. To connect the physical room to the DTS, the Internet of Things (IoT) devices installed within the room must be linked to the digital system through an appropriate network connection. This connection enables the devices to transmit data continuously to the DTS, where the data can be processed and used for monitoring, diagnosis, and control. For this process to work effectively, the connection between the physical room and the DTS must be both secure and reliable. If the connection is unstable, inaccurate, or vulnerable to disruption, the value of the entire system will be compromised. Figure 6.4 summarises this third framework component, which focuses on the connection to the digital system.

The first issue to consider is the **compatibility of the IoT devices**. When setting up the DTS, ensure the selected IoT devices are compatible with the system. This helps to prevent software, hardware, or communication protocol conflicts that may weaken system performance or disrupt data transfer (Jimenez, Jahankhani and Kendzierskyj, 2020). Closely related to this is the **deployment of IoT devices**. A clear plan should be developed for the installation, positioning, and testing of the devices to ensure proper, efficient deployment. In the context of IEQ management, this is especially important because sensors must be placed in those parts of the room where the monitored environmental conditions are most critical (Cai et al., 2017).

Another major consideration is establishing a **reliable and secure communication protocol**. A strong communication protocol is necessary to ensure that

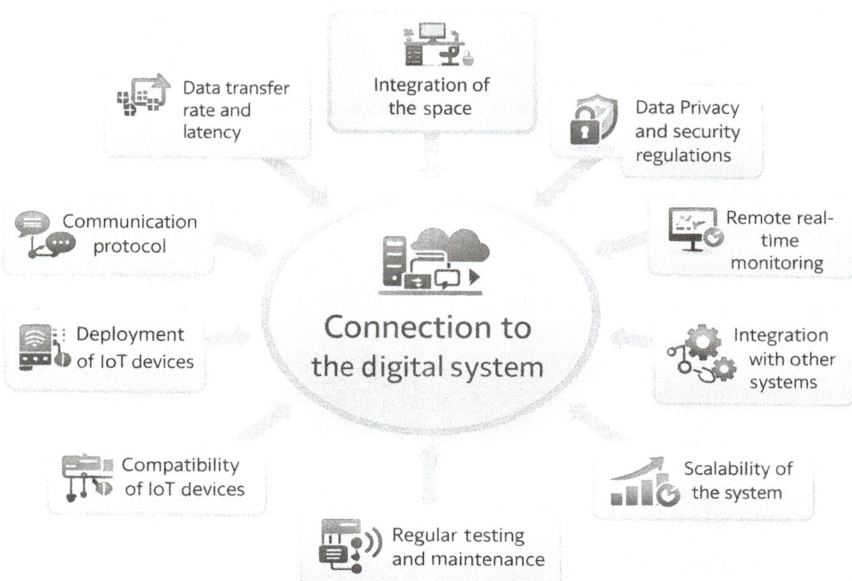

Figure 6.4 Framework component 3: Connection to the digital system.

data moves between the physical room and the DTS without breach, distortion, or loss of information (Jimenez, Jahankhani and Kendzierskyj, 2020). Alongside this, the **data transfer rate and latency** must also be considered. Real-time monitoring depends on the system's ability to move data quickly and with minimal delay. If transfer rates are too slow, or if latency is too high, the DTS will not be able to represent the physical room accurately in real time (Khajavi et al., 2019; Ma et al., 2019). This would weaken the system's responsiveness and reduce its usefulness for timely decision-making.

The **integration of the space** into the DTS must also be addressed carefully. This means that the existing infrastructure and operational features of the physical room must be aligned with the characteristics and functions of the digital system (Kousi et al., 2019; Tao, Zhang and Nee, 2019). In addition, **data privacy and security regulations** must be respected. Since the DTS deals with environmental, operational, and system-related information, it is important to ensure the system complies with relevant data privacy and security requirements to prevent unauthorised access and maintain trust in the system (Lontsikh et al., 2020; Shahzad et al., 2022).

A further issue is the need for **remote real-time monitoring**. The DTS should enable remote monitoring of the physical room via a live digital representation. This can take the form of audio, visual, or informational representation, depending on the specific needs of the system and the users (Hu and Li, 2013; Zhu, Liu and Xu, 2019). The system should also support **integration with other systems** to obtain a broader, more comprehensive view of the indoor environment. This may include integration with building systems, operational databases, and BIM-related information, thereby bringing data together in a single, coordinated environment for easier monitoring and analysis (Alonso et al., 2019; Tchana, Ducellier and Remy, 2019; Hosamo et al., 2023).

Finally, the framework must account for the **system's scalability** and the need for **regular testing and maintenance**. Scalability is important because the DTS should be capable of expanding in the future to accommodate more rooms, additional sensors, or even entire buildings (Uhlemann et al., 2017b; Kan and Anumba, 2019). At the same time, regular testing and maintenance are essential to ensure that the system and its components remain effective and reliable in measuring data and supporting corrective action. This helps to ensure that the IoT devices and the DTS remain in proper working condition over time (Khajavi et al., 2019; Yu et al., 2022).

These conditions collectively help establish a seamless and dependable connection between the physical room and the DTS. This connection is central to achieving accurate, secure, and real-time monitoring within the IEQ management framework.

6.5 Framework Component 4: Digital Rendering and Display

Once the connection to the DTS has been established, the next step is to render and display the information in a user-friendly and meaningful way. At this stage, the value of the collected data depends not only on its accuracy but also on how clearly it can be presented to the user. This is achieved through a virtual reality (VR) engine and a graphical user interface (GUI) integrated with the DTS. Together, the VR engine and GUI provide an intuitive and accessible interface through which users can visualise, interpret, and interact with the information collected from the IoT devices. This includes not only digital representations of the monitored space but also graphs, charts, and other visual tools that help users identify patterns, trends, and anomalies more easily.

The DTS should also allow users to interact with the data in real time. This can be done by adjusting display settings, applying filters, or selecting specific parameters to focus on. Such functionality enables users to quickly identify important information, especially when data falls outside the acceptable range or threshold values defined in the framework. In addition, the DTS should be optimised for use across different device types and operating systems, such as Windows OS and Mac OS, so that users can access the system flexibly and remotely whenever needed. Figure 6.5 summarises this fourth framework component, which focuses on digital rendering and display.

The first consideration is **display compatibility**. It is essential to ensure that the display interface is compatible with different devices, operating systems, and screen sizes. The information display system should therefore function effectively across a wide range of platforms in order to maximise accessibility and user reach (Um, Weyer and Quint, 2017). The second consideration is a **customisable display**. The system should allow users to customise which information is shown and how it is presented. This is important because different users may have different monitoring priorities and preferences (Mateev, 2020).

The third consideration is **ease of navigation**. The display system should have an easy-to-use interface that allows users to quickly access relevant information. This requires clear layout design, intuitive navigation tools, and a logical arrangement of displayed content (Kousi et al., 2019). The fourth

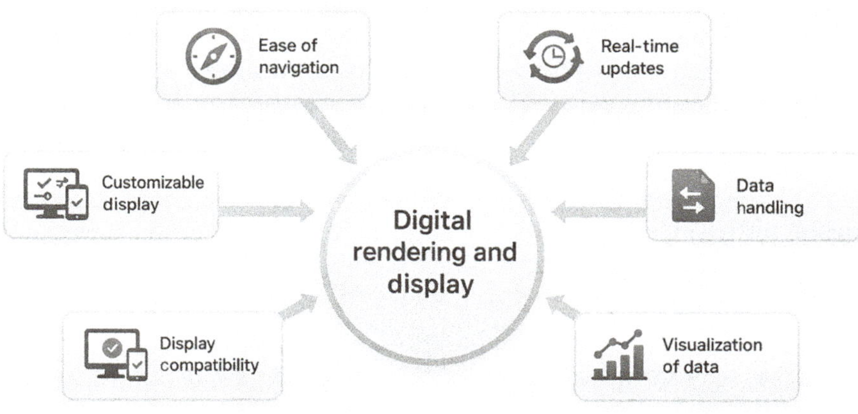

Figure 6.5 Framework component 4: Digital rendering and display.

consideration is **real-time updating**. The information presented on the screen must be continuously updated so that the displayed conditions always reflect the most current state of the physical environment. Provision should also be made for remote monitoring of changes triggered by IoT devices (Uhlemann et al., 2017b).

The fifth consideration is **data visualisation**. The display system should be capable of displaying data in different visual formats, including live graphs, charts, dashboards, and graphical renderings. Good visualisation supports better understanding and analysis by helping users identify trends and patterns more quickly (Dawkins, Hudson-Smith and Dennett, 2018; Lutters and Damgrave, 2019). The sixth consideration is **data handling**. The information display system should be able to handle large volumes of data without compromising system responsiveness or performance. This is important because the DTS may process continuous streams of data over time, and the system must remain stable and efficient even under heavy data load (Kaewunruen and Xu, 2018).

By carefully considering these conditions, a digital rendering and display system can be developed to present information from IoT sensors clearly and effectively. This strengthens the usability of the DTS and supports accurate, accessible, and real-time monitoring for IEQ management.

6.6 Framework Component 5: Identification of Issues

The fifth component of the DTS framework is the identification of possible issues. This is a critical stage because the value of monitoring lies not only in collecting data but also in using it to recognise when indoor environmental conditions deviate from acceptable performance. In the DTS, this is achieved by analysing data that falls within and outside the established threshold ranges. For the system to function efficiently, it must be able to distinguish between normal conditions and conditions that suggest a developing or existing IEQ problem. One of the most practical ways to achieve this is to define upper and lower threshold values for each monitored parameter and then compare live data against those limits. When data falls outside the acceptable range, the system may flag it as a potential IEQ issue requiring attention.

The ability to identify issues promptly is central to the DTS's purpose. Once a potential problem has been detected, the system can trigger immediate responses and send notifications to the relevant personnel. This ensures that problems are not ignored or discovered too late. Instead, action can be taken in a timely and organised manner, thereby supporting smoother operation, better indoor environmental control, and more reliable system performance. Figure 6.6 summarises this fifth framework component, which focuses on identifying issues.

The first important condition is the **integration of thresholds**. The system must be able to incorporate clearly defined upper and lower limits for the collected data. Any reading beyond these limits should be interpreted as out of

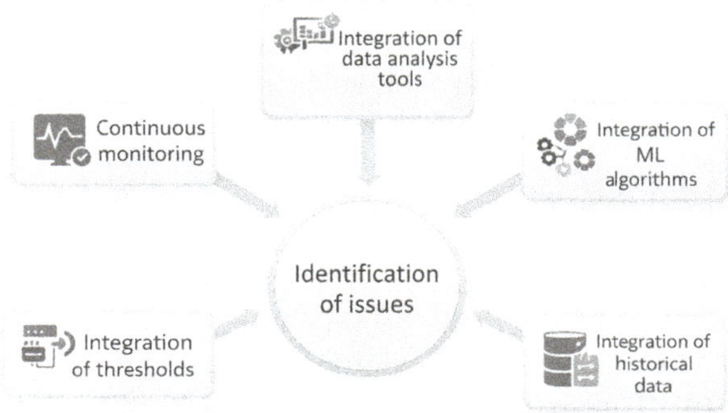

Figure 6.6 Framework component 5: Identification of issues.

range. This information is essential because it helps determine the condition, health, and performance of the physical room being monitored (Tao et al., 2018; Jimenez, Jahankhani and Kendzierskyj, 2020). Without properly integrated thresholds, the DTS would not be able to recognise when environmental conditions have become unacceptable.

The second condition is **continuous monitoring**. The system should be designed to monitor incoming data in real time so that deviations from established thresholds can be detected as soon as they occur. This supports quick issue identification and reduces the delay between the emergence of a problem and the system's recognition of that problem (Shim et al., 2019; Bende and Gochhait, 2020; Singh et al., 2022). In the context of IEQ management, such responsiveness is especially important because indoor conditions can change quickly and may begin to affect occupant comfort or health if not addressed promptly.

The third condition is the **integration of data analysis tools**. The system should include built-in analytical tools to help identify the likely cause of any deviation from the established threshold. This adds depth to the monitoring process by moving the system beyond simple detection and towards interpretation. Once the probable cause of an issue has been identified, the information can be used to support faster and more accurate resolution (Mateev, 2020).

The fourth condition is the **integration of machine learning algorithms**. The system should be designed to enable machine learning methods to identify patterns in the data and predict issues before they fully develop. This creates a more proactive system, one that not only responds to problems after they occur but also anticipates them. Such capability can reduce downtime, improve system stability, and strengthen the effectiveness of the DTS (Min et al., 2019; Kaur, Mishra and Maheshwari, 2020).

The fifth condition is the **integration of historical data**. The system should be able to store, retrieve, and analyse past data in order to identify recurring trends and patterns. Historical data provides valuable context by helping distinguish isolated events from recurring or emerging issues. It also allows the

DTS to learn from past system behaviour and improve its ability to identify potential problems before they occur (Gao and Pishdad-Bozorgi, 2019; Qi et al., 2019).

By carefully considering these conditions, the DTS can identify potential issues more effectively and provide the information needed for prompt, accurate intervention. In this way, issue identification becomes not merely a technical function, but a key decision-support stage within the broader framework for IEQ management.

6.7 Framework Component 6: Selection of Corrective Measures

The sixth component of the DTS framework involves selecting corrective measures for the identified issues. Once the system has detected a potential problem through data collection and threshold-based analysis, the next step is to determine the most appropriate response. This stage is important because issue identification alone does not improve IEQ. The value of the DTS lies in its ability to move from detection to action in a timely and informed manner. The corrective measure selected must reflect the nature, cause, and severity of the issue, as interpreted from the data collected and analysed against the established thresholds. For example, if the system detects that the room's ambient temperature falls outside the acceptable range, the corrective measure should focus on adjusting the room's thermal conditions accordingly.

In some situations, the corrective measure may involve an automated action that directly addresses the condition that has exceeded the defined range or threshold. These actions are usually preprogrammed into the DTS design algorithm and are triggered automatically when specific conditions are met. In other situations, the corrective measure may involve an automated notification sent to a designated person, who must then initiate a more suitable response or further intervention. At the end of each data analysis cycle, the DTS should therefore be able to select the most appropriate action for the detected condition. This selected action is then transmitted from the system's backend for implementation in the physical building. Whether the response is automatic or human assisted, it is important that the decision is made quickly and effectively to minimise the impact of the identified issue. Figure 6.7 summarises this sixth framework component, which focuses on selecting corrective measures.

One of the first conditions to consider in this stage is the **root cause** of the issue. Confirming and understanding the root cause is essential if the problem is to be addressed properly. This ensures that the selected corrective measure targets the actual source of the problem rather than merely responding to its visible symptoms. In this sense, the issue identified in the previous framework component is examined more critically here to support a more accurate and meaningful intervention (Lydon et al., 2019).

Another important condition is the availability of **multiple corrective measures**. The system should not rely on a single response option. Instead, it should be able to evaluate several possible actions and select the one that best addresses

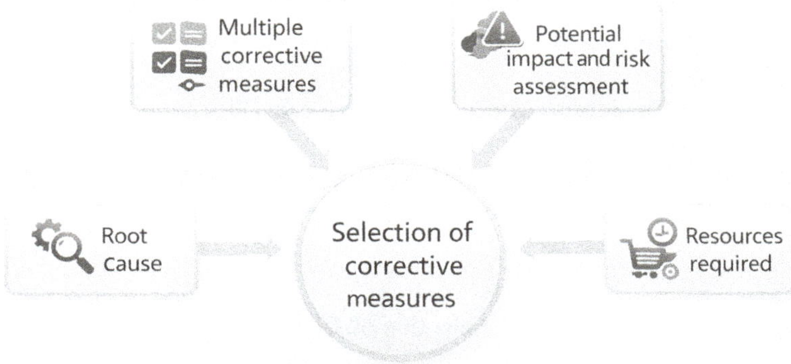

Figure 6.7 Framework component 6: Selection of corrective measures.

the problem while also considering the system's broader conditions and constraints (Uhlemann, Lehmann and Steinhilper, 2017a; Fotland, Haskins and Rølvåg, 2020). This gives the DTS greater flexibility and improves its ability to respond intelligently to different IEQ situations.

It is also necessary to consider the **potential impact and risk assessment**. Before any corrective measure is implemented, the possible effect of that action on the wider system, including the physical room, should be assessed. Every intervention carries the possibility of creating new conditions or unintended consequences. Evaluating the risks associated with each response option helps reduce the likelihood of introducing new problems while addressing existing ones (Damjanovic-Behrendt, 2018). This makes the selection process more careful, balanced, and reliable.

Another condition is the **resources required** for implementation. Corrective measures often depend on the availability and capability of IoT devices, control systems, or other technical components needed to effect the required change. For this reason, it is important to consider the resources required for each possible response to assess the feasibility of implementation (Uhlemann et al., 2017b). A corrective measure may be appropriate in principle but ineffective in practice if the necessary devices or system functions are unavailable at the required time.

By carefully considering these conditions, the DTS can select corrective measures more effectively for the issues it identifies. This strengthens the framework by ensuring that responses are not only fast but also appropriate, feasible, and aligned with the broader goal of maintaining healthy, comfortable IEQ conditions in the indoor environment.

6.8 Framework Component 7: Implementation of Corrective Measures

The seventh and final component of the framework for developing a DTS is the implementation of corrective measures. This is the stage at which the decisions reached in the earlier steps are translated into action within the physical environment. At this point, the value of the system becomes visible because the DTS no longer only monitors, analyses, and selects responses. It also enables those responses to be carried out in the room or building. In this framework, corrective measures are implemented primarily through IoT devices, especially actuators that can receive instructions from the backend system and respond accordingly.

The IoT devices should receive signals from the backend system and act in line with the selected response. In this context, the devices are used to address issues identified in earlier steps of framework. This stage is important because it brings the real-time monitoring system into practical operation. The required actions are predefined and automated as part of the DTS design algorithm. Once a suitable corrective action has been selected, that pre-programmed instruction is transmitted to the IoT devices installed within the building so that the necessary adjustment can be made. Figure 6.8 summarises the seventh framework component, which focuses on implementing corrective measures.

One important condition at this stage is the **provision of automatic controls**. The choice of IoT devices depends on the type of corrective measure that needs to be implemented. The selected device must be capable of carrying out the required action effectively. Automatic controls should therefore be built into the actuators based on the analysed DTS data. This helps ensure that corrective measures are implemented quickly, accurately, and consistently whenever the system detects a need for intervention.

A second condition is the **implementation of manual controls**. Even though automation is central to the DTS, manual override controls should also be provided. These are necessary as a backup in situations where automatic controls fail, produce errors, or prove inadequate for the condition being managed. Manual intervention also remains important in cases where human judgment is required. The inclusion of manual controls, therefore, strengthens the system's reliability and flexibility.

A third condition is **automated notification**. In addition to direct automatic control, the DTS should be able to send alerts to designated persons, such as

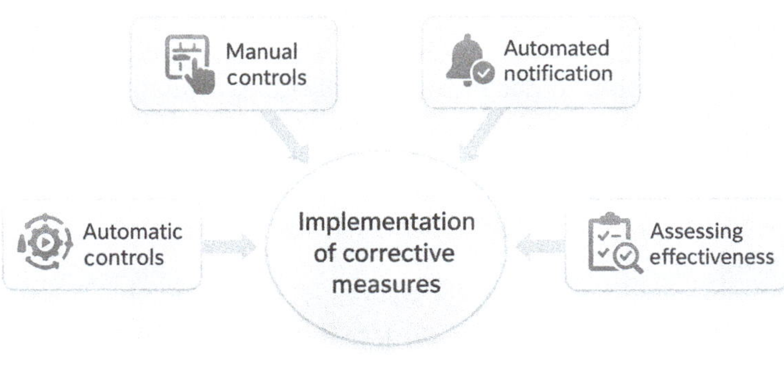

Figure 6.8 Framework component 7: Implementation of corrective measures.

facility managers, whenever monitored data falls outside the established threshold. This enables responsible personnel to take further action as needed. The system should also be able to trigger external notification mechanisms, such as email alerts, so that a response can begin immediately, even when direct physical intervention is required.

A fourth condition is the need to **monitor effectiveness**. After a corrective measure has been implemented, the DTS should continue to monitor the affected conditions and assess whether the action has successfully resolved the issue. If the measure is not fully effective, the system should support further adjustments or escalation. This is important because implementation is not complete until the action's effectiveness has been confirmed.

By carefully considering these conditions, the DTS can implement corrective measures effectively and reliably in response to identified issues. This final framework component completes the system's operational logic by ensuring that insights from monitoring and analysis translate into practical actions that support healthy, comfortable IEQ conditions.

6.9 Logic of the Digital Twin System Framework for IEQ Management

The logic of the DTS framework for IEQ management is based on a connected, sequential process in which indoor environmental conditions are sensed, interpreted, evaluated, and corrected in a continuous cycle. The framework is designed to move beyond isolated monitoring by establishing a structured relationship between the physical indoor environment and its digital representation. In this way, the DTS functions not merely as a data collection tool but as an intelligent management system that supports timely and informed intervention. The underlying logic of the framework is that each component depends on the one before it while also setting the conditions for the next. Together, the components form a complete pathway from environmental observation to corrective action.

The process starts with identifying the necessary data and devices. This initial step clarifies what needs to be measured, why it is important, and which sensors and actuators are required for that purpose. Without this foundation, the system cannot function effectively. After determining the relevant data and devices, the next step is to establish acceptable ranges and thresholds for each IEQ parameter. This provides the DTS with a standard to evaluate current conditions. The framework then connects to the digital system, linking the physical environment to the DTS via IoT devices and network communication. This connection allows real-time data exchange between the room and the digital platform.

After connecting the system, the data must be presented in a way that is understandable and useful to the user. This is the role of digital rendering and display, where the DTS transforms raw data into visual and interactive information through dashboards, graphical displays, and digital representations.

Once displayed, the data can then be interpreted more effectively. The next step is to identify issues where the system compares incoming data against predefined thresholds to detect deviations that may indicate an IEQ problem. This stage is critical because it translates environmental data into actionable knowledge.

Following issue identification, the framework proceeds to select corrective measures. At this point, the DTS decides on the most appropriate response based on the type, severity, and likely cause of the problem. The final stage involves implementing the corrective measures, in which the chosen response is executed either automatically via IoT actuators or via notifications to designated personnel for manual intervention. The framework thus closes the loop by ensuring that system intelligence results in physical action within the monitored environment.

The logic of the DTS framework is therefore cyclical, responsive, and adaptive. Each cycle of sensing, analysis, evaluation, and action improves the system's ability to manage IEQ conditions effectively. In this way, the framework provides a practical structure for achieving healthier, more comfortable, and better-controlled indoor environments through DT-enabled IEQ management.

6.10 Summary

This chapter presented the DTS framework for IEQ management and explained the logic through which the system operates. The discussion showed that effective IEQ management requires more than simple monitoring. It requires a structured process that can capture indoor environmental data, interpret changing conditions, identify developing issues, and support timely corrective action. In response to this need, the chapter introduced a framework comprising seven interconnected components.

The framework begins with identifying the required data and devices, followed by the establishing acceptable ranges and thresholds for the monitored parameters. It then addresses the connection between the physical space and the digital system, and the rendering of the collected information in a form that is accessible and meaningful to users. The latter components focus on issue identification, the selection of corrective measures, and the implementation of those measures within the physical environment. Together, these stages form a connected and responsive system for managing IEQ more effectively.

Ultimately, the chapter showed that the DTS framework provides a practical and intelligent structure for real-time IEQ management. It supports healthier, more comfortable, and better-controlled indoor environments through continuous monitoring, analysis, and action.

References

Alonso, R. et al. (2019) 'SPHERE: BIM Digital Twin platform', *Proceedings*, 20(9), pp. 1–5. doi:10.3390/proceedings2019020009

Bende, A. and Gochhait, S. (2020) 'Leveraging Digital Twin technology in the healthcare industry – A machine learning based approach', *European Journal of Molecular & Clinical Medicine*, 24(2), pp. 5592–5602. Available at: https://www.psychosocial.com/index.php/ijpr/article/view/2388/2206 (Accessed: 9 April 2023).

Botkina, D. et al. (2018) 'Digital Twin of a cutting tool', in *Proceedings of the 51st CIRP Conference on Manufacturing Systems*. Elsevier, pp. 215–218. doi:10.1016/j.procir.2018.03.178

BSI Group (2019) Digital twins for the built environment. *BIMToday*. Available at: https://www.pbctoday.co.uk/news/bim-news/digital-twin-4-0/64519

Cai, J. et al. (2023) 'Digital twin for healthy indoor environment: A vision for the post-pandemic era', *Frontiers of Engineering Management*, pp. 1–19. doi:10.1007/S42524-022-0244-Y

Cai, Y. et al. (2017) 'Sensor data and information fusion to construct digital-twins virtual machine tools for cyber-physical manufacturing', *Procedia Manufacturing*, 10, pp. 1031–1042. doi:10.1016/j.promfg.2017.07.094

Damjanovic-Behrendt, V. (2018) 'A digital twin-based privacy enhancement mechanism for the automotive industry', in *Proceedings of the 9th International Conference on Intelligent Systems: Theory, Research and Innovation in Applications*, pp. 272–279. doi:10.1109/IS.2018.8710526

Dawkins, O., Hudson-Smith, A. and Dennett, A. (2018) 'Living with a Digital Twin: Operational management and engagement using IoT and mixed realities at UCL's here east campus on the Queen Elizabeth Olympic Park', *BIM-DigiCons*, p. 6. Available at: https://pdfs.semanticscholar.org/da61/a1bf345892d7504bde14624788ee9dd07b47.pdf (Accessed: 7 January 2020).

Fotland, G., Haskins, C. and Rølvåg, T. (2020) 'Trade study to select best alternative for cable and pulley simulation for cranes on offshore vessels', *Systems Engineering*, 23(2), pp. 177–188. doi:10.1002/SYS.21503

Ganguli, R. and Adhikari, S. (2020) 'The digital twin of discrete dynamic systems: Initial approaches and future challenges', *Applied Mathematical Modelling*, 77, pp. 1110–1128. doi:10.1016/j.apm.2019.09.036

Gao, X. and Pishdad-Bozorgi, P. (2019) 'A framework of developing machine learning models for facility life-cycle cost analysis', *Building Research and Information*, pp. 1–25. doi:10.1080/09613218.2019.1691488

Haag, S. and Anderl, R. (2018) 'Digital twin – Proof of concept', *Manufacturing Letters*, 15, pp. 64–66. doi:10.1016/j.mfglet.2018.02.006

Haag, S. and Anderl, R. (2019) 'Automated Generation of as-manufactured geometric representations for digital twins using STEP', in *Proceedings of the 29th CIRP Design*. Elsevier B.V., pp. 1082–1087. doi:10.1016/j.procir.2019.04.305

Hosamo, H.H. et al. (2023) 'Digital Twin framework for automated fault source detection and prediction for comfort performance evaluation of existing non-residential Norwegian buildings', *Energy and Buildings*, 281, p. 112732. doi:10.1016/J.ENBUILD.2022.112732

Hu, Q. and Li, F. (2013) 'Hardware design of smart home energy management system with dynamic price response', *IEEE Transactions on Smart Grid*, 4(4), pp. 1878–1887. doi:10.1109/TSG.2013.2258181

Jimenez, J.I., Jahankhani, H. and Kendzierskyj, S. (2020) 'Health care in the cyberspace: Medical cyber-physical system and digital twin challenges', *Internet of Things*, pp. 79–92. doi:10.1007/978-3-030-18732-3_6

Kaewunruen, S. and Xu, N. (2018) 'Digital twin for sustainability evaluation of railway station buildings', *Frontiers in Built Environment*, 4, p. 77. doi:10.3389/fbuil.2018.00077

Kan, C. and Anumba, C.J. (2019) 'Digital Twins as the next phase of cyber-physical systems in construction', in *Proceedings of the ASCE International Conference on Computing in Civil Engineering*. Atlanta: American Society of Civil Engineers (ASCE), pp. 256–264. doi:10.1061/9780784482438.033

Kaur, M.J., Mishra, V.P. and Maheshwari, P. (2020) 'The convergence of Digital Twin, IoT, and machine learning: transforming data into action', in M. Farsi, A. Daneshkhah, A. Hosseinian-Far and H. Jahankhani (eds.), *Digital Twin technologies and smart cities*. Cham, Switzerland: Springer, pp. 3–17. doi:10.1007/978-3-030-18732-3_1

Khajavi, S.H. et al. (2019) 'Digital Twin: Vision, benefits, boundaries, and creation for buildings', *IEEE Access*, 7. doi:10.1109/ACCESS.2019.2946515

Kousi, N. et al. (2019) 'Digital Twin for adaptation of robots' behavior in flexible robotic assembly lines', in *Proceedings of the International Conference on Changeable, Agile, Reconfigurable and Virtual Production*. Elsevier, pp. 121–126. doi:10.1016/j.promfg.2018.12.020

Lontsikh, P.A. et al. (2020) 'Information security methods' application based on the digital management approaches and the Deming cycle in improving the modern production's processes', in *Proceedings of the IEEE International Conference "Quality Management, Transport and Information Security, Information Technologies" (IT and QM and IS)*, pp. 123–126. doi:10.1109/ITQMIS51053.2020.9322856

Lu, Y. et al. (2020) 'Digital Twin-driven smart manufacturing: Connotation, reference model, applications and research issues', *Robotics and Computer-Integrated Manufacturing*, p. 101837. doi:10.1016/j.rcim.2019.101837

Lutters, E. and Damgrave, R. (2019) 'The development of pilot production environments based on digital twins and virtual dashboards', in *Proceedings of the 29th CIRP Design*. Elsevier, pp. 94–99. doi:10.1016/j.procir.2019.04.228

Lydon, G.P. et al. (2019) 'Coupled simulation of thermally active building systems to support a digital twin', *Energy and Buildings*, 202, p. 109298. doi:10.1016/j.enbuild.2019.07.015

Ma, X. et al. (2019) 'Digital twin enhanced human-machine interaction in product lifecycle', in *Proceedings of the 11th CIRP Conference on Industrial Product-Service Systems*. Elsevier, pp. 789–793. doi:10.1016/j.procir.2019.04.330

Mateev, M. (2020) 'Industry 4.0 and the Digital Twin for building industry', *International Scientific Journal "Industry 4.0"*, 5(1), pp. 29–32. Available at: https://stumejournals.com/journals/i4/2020/1/29.full.pdf (Accessed: 7 January 2020).

Min, Q. et al. (2019) 'Machine learning based Digital Twin framework for production optimization in petrochemical industry', *International Journal of Information Management*, 49, pp. 502–519. doi:10.1016/j.ijinfomgt.2019.05.020

Murphy, A. et al. (2020) 'Representing financial data streams in digital simulations to support data flow design for a future Digital Twin', *Robotics and Computer-Integrated Manufacturing*, 61. doi:10.1016/j.rcim.2019.101853

Qi, Q. et al. (2019) 'Enabling technologies and tools for digital twin', *Journal of Manufacturing Systems*. doi:10.1016/j.jmsy.2019.10.001

Re Cecconi, F. et al. (2021) 'Soft Digital Twin for IEQ enabling the COVID risk mitigation in educational spaces', pp. 1–12. Available at: https://re.public.polimi.it/handle/11311/1187733 (Accessed: 30 April 2023).

Shahzad, M. et al. (2022) 'Digital Twins in built environments: An investigation of the characteristics, applications, and challenges', *Buildings*, 12(2), p. 120. doi:10.3390/BUILDINGS12020120

Shim, C.S. et al. (2019) 'Development of a bridge maintenance system for prestressed concrete bridges using 3D digital twin model', *Structure and Infrastructure Engineering*, 15(10), pp. 1319–1332. doi:10.1080/15732479.2019.1620789

Singh, M. et al. (2022) 'Applications of Digital Twin across industries: A review', *Applied Sciences*, 12(11), p. 5727. doi:10.3390/APP12115727

Tao, F., Zhang, M. and Nee, A.Y.C. (2019) 'Digital Twin and virtual reality and augmented reality/mixed reality', *Digital Twin Driven Smart Manufacturing*, pp. 219–241. doi:10.1016/B978-0-12-817630-6.00011-4

Tao, F. et al. (2018) 'Digital twin driven prognostics and health management for complex equipment', *CIRP Annals - Manufacturing Technology*, 67(1), pp. 169–172. doi:10.1016/j.cirp.2018.04.055

Tchana, Y., Ducellier, G. and Remy, S. (2019) 'Designing a unique Digital Twin for linear infrastructure lifecycle management', *Procedia CIRP*, 84, pp. 545–549. doi:10.1016/j.procir.2019.04.176

Uhlemann, T.H.J., Lehmann, C. and Steinhilper, R. (2017a) 'The Digital Twin: Realizing the cyber-physical production system for Industry 4.0', in *Proceedings of the 24th CIRP Conference on Life Cycle Engineering*, pp. 335–340. doi:10.1016/j.procir.2016.11.152

Uhlemann, T.H.J. et al. (2017b) 'The Digital Twin: Demonstrating the potential of real time data acquisition in production systems', *Procedia Manufacturing*, 9, pp. 113–120. doi:10.1016/j.promfg.2017.04.043

Um, J., Weyer, S. and Quint, F. (2017) 'Plug-and-simulate within modular assembly line enabled by Digital Twins and the use of AutomationML', *IFAC-PapersOnLine*, 50(1), pp. 15904–15909. doi:10.1016/j.ifacol.2017.08.2360

van Midden, J.N.A. (2022) Linked data-based Digital Twins for increasing the Indoor Environmental Quality (IEQ) of individuals working from home [Master's dissertation]. Eindhoven University of Technology. Available at: https://research.tue.nl/en/studentTheses/linked-data-based-digital-twins-for-increasing-the-indoor-environ (Accessed: 30 April 2023).

Vathoopan, M. et al. (2018) 'Modular fault ascription and corrective maintenance using a Digital Twin', *IFAC-PapersOnLine*, 51(11), pp. 1041–1046. doi:10.1016/j.ifacol.2018.08.470

Woodhead, R., Stephenson, P. and Morrey, D. (2018) 'Digital construction: From point solutions to IoT ecosystem', *Automation in Construction*, 93, pp. 35–46. doi:10.1016/j.autcon.2018.05.004

Yu, W. et al. (2022) 'Energy digital twin technology for industrial energy management: Classification, challenges and future', *Renewable and Sustainable Energy Reviews*, 161, p. 112407. doi:10.1016/J.RSER.2022.112407

Zhu, Z., Liu, C. and Xu, X. (2019) 'Visualisation of the digital twin data in manufacturing by using augmented reality', *Procedia CIRP*, 81, pp. 898–903. doi:10.1016/j.procir.2019.03.223

Index

Pages in *italics* refer to figures and pages in **bold** refer to tables

For Product Safety Concerns and Information please contact our EU representative GPSR@taylorandfrancis.com
Taylor & Francis Verlag GmbH, Kaufingerstraße 24, 80331 München, Germany

www.ingramcontent.com/pod-product-compliance
Ingram Content Group UK Ltd.
Pitfield, Milton Keynes, MK11 3LW, UK
UKHW021027031026
476244UK00003B/42

* 9 7 8 1 0 4 1 2 3 3 8 9 3 *